Lecture Notes in Electrical Engineering

Volume 606

The book series *Lecture Notes in Electrical Engineering* (LNEE) publishes the latest developments in Electrical Engineering - quickly, informally and in high quality. While original research reported in proceedings and monographs has traditionally formed the core of LNEE, we also encourage authors to submit books devoted to supporting student education and professional training in the various fields and applications areas of electrical engineering. The series cover classical and emerging topics concerning:

- Communication Engineering, Information Theory and Networks
- Electronics Engineering and Microelectronics
- Signal, Image and Speech Processing
- Wireless and Mobile Communication
- Circuits and Systems
- Energy Systems, Power Electronics and Electrical Machines
- Electro-optical Engineering
- Instrumentation Engineering
- Avionics Engineering
- Control Systems
- Internet-of-Things and Cybersecurity
- Biomedical Devices, MEMS and NEMS

For general information about this book series, comments or suggestions, please contact leontina.dicecco@springer.com.

To submit a proposal or request further information, please contact the Publishing Editor in your country:

China
Jasmine Dou, Associate Editor (jasmine.dou@springer.com)

India
Swati Meherishi, Executive Editor (swati.meherishi@springer.com)
Aninda Bose, Senior Editor (aninda.bose@springer.com)

Japan
Takeyuki Yonezawa, Editorial Director (takeyuki.yonezawa@springer.com)

South Korea
Smith (Ahram) Chae, Editor (smith.chae@springer.com)

Southeast Asia
Ramesh Nath Premnath, Editor (ramesh.premnath@springer.com)

USA, Canada:
Michael Luby, Senior Editor (michael.luby@springer.com)

All other Countries:
Leontina Di Cecco, Senior Editor (leontina.dicecco@springer.com)
Christoph Baumann, Executive Editor (christoph.baumann@springer.com)

**** Indexing: The books of this series are submitted to ISI Proceedings, EI-Compendex, SCOPUS, MetaPress, Web of Science and Springerlink ****

More information about this series at http://www.springer.com/series/7818

Eric Todd Quinto • Nathan Ida • Ming Jiang
Alfred K. Louis
Editors

The Proceedings of the International Conference on Sensing and Imaging, 2018

Springer

Editors
Eric Todd Quinto
Department of Mathematics
Tufts University
Medford, MA, USA

Nathan Ida
Department of Electrical & Computer Engineering
The University of Akron
Akron, OH, USA

Ming Jiang
Department of Information Science
School of Mathematical Sciences
Peking University
Beijing, China

Alfred K. Louis
Mathematik
Univ. des Saarlandes
Saarbrücken, Saarland, Germany

ISSN 1876-1100 ISSN 1876-1119 (electronic)
Lecture Notes in Electrical Engineering
ISBN 978-3-030-30824-7 ISBN 978-3-030-30825-4 (eBook)
https://doi.org/10.1007/978-3-030-30825-4

This Springer imprint is published by the registered company Springer Nature Switzerland AG.
The registered company address is: Gewerbestrasse 11, 6330 Cham, Switzerland

Foreword

Simin Li

The International Conference on Sensing and Imaging (ICSI) is an important academic conference that provides an excellent platform for international exchange and collaboration in various fields of sensing and imaging such as image detection, image reconstruction, as well as industrial and biomedical applications. Guangxi University of Science and Technology (GXUST) was honored to be the host of the ICSI 2018 held at Liuzhou, Guangxi, China, from October 15 to 17, 2018.

GXUST is rooted in its two major components and predecessors: Guangxi College of Engineering founded in 1958 and Medical College of Liuzhou founded in 1951. The mission of the University is to cultivate talent and innovative spirit in engineering and medical science students.

To ensure the success of the conference and promote interdisciplinary collaboration, we invited domestic and overseas experts and scholars specializing in sensing and imaging to gather here at Liuzhou to exchange ideas on hot topics and cutting-edge research in sensing and imaging, such as image reconstruction, image restoration, image fusion, biomedical imaging, nondestructive inspection, and remote sensing. We sincerely hope that theoretical and practical research in sensing and imaging can be continuously promoted through academic exchanges, technical innovation, and cooperation after this conference.

I would like to thank the co-chairs of the conference, Professors Ming Jiang of Peking University and Nathan Ida of the University of Akron, for their tireless efforts to initiate the conference in a brand new location with a new host, GXUST, and to co-chair the conference. I would also like to thank Prof. Eric Todd Quinto of Tufts University for leading the effort to edit and publish the conference proceeding and Professor Yu Shang of North University China for a special issue of *Sensing and Imaging*, with a strong editorial and administrative support from Professors Ming Jiang, Nathan Ida, Alfred K Louis of Saarland University, Yu Shang of the North University of China, Tobias Kluth of the University of Bremen, and Hong

---

S. Li
Guangxi University of Science and Technology (GXUST), Liuzhou, China

Liu of GXUST. My special thanks go to all the local organizers and volunteers, especially Prof. Hong Liu and his team at GXUST, for their steadfast efforts to make this conference happen. Finally, I thank all the invited speakers, session chairs, and participants of this conference. It is the joint effort of all the aforementioned colleagues that made this conference a successful one.

Preface

The International Conference on Sensing and Imaging 2018 (ICSI 2018) was the fourth conference in this series. The previous three were held at Chifeng, Inner Mongolia, China, in 2015; Taiyuan, Shanxi Province, China, in 2016; and at Chengdu, Sichuan Province, China, in 2017.

This fourth conference was held in Liuzhou, China, from October 15 to 17, 2018. Guangxi University of Science and Technology (GXUST) was the primary sponsor of ICSI 2018. Prof. Simin Li, the President of GXUST, was the general chair for this conference. Profs. Nathan Ida and Ming Jiang were co-chairs. We accepted 58 submissions for presentation at the conference. With the success of the previous conferences and because of suggestions from peers and requests from participants, we decided to publish the proceedings for ICSI 2018.

These proceedings have 17 papers including 8 invited chapters. The review and publication of these proceedings are not like most conference proceedings. All submissions were first reviewed to see if they could be suitable for presentations at the conference and were accepted for presentation after revisions. Unlike other conference proceedings, after the conference, all revised submissions underwent another round of revision so the authors could incorporate discussions at the conference. They were reviewed again for final acceptance in the proceedings, and some submissions were rejected. We understand that proceedings are different from journal submissions, but we hope the two rounds of revision improve the quality of submissions without delaying publication.

Seven additional submissions were selected to be published in the special issue "Recent developments in Sensing and Imaging" in the journal *Sensing and Imaging* published by Springer Nature. Each selected manuscript has been extended to a full-length regular paper with significant additions from its conference version and reviewed under the normal reviewing process of that journal. This special issue is edited by lead guest editor, Yu Shang North University of China, China, and guest editors Tobias Kluth, University of Bremen, Germany, and Hong Liu, Guangxi University of Science and Technology.

We thank the publishing editor, Christopher T. Coughlin of Springer Nature, for his prompt agreement and coordination to publish the proceedings in Lecture Notes in Electrical Engineering (LNEE), so that authors could prepare their manuscript for this publication opportunity. We thank all the reviewers for their timely responses and comments for authors to improve the quality of manuscripts. We thank all the authors for their participation in ICSI 2018 and their understanding and patience for the two rounds of revisions. We thank authors of the 8 invited contributions for their effort in preparing the invited chapters. Finally, we thank, Chandhini Kuppusamy, Ho Yin Fan, and Jeffrey Taub, of the editorial team of Springer Nature for their efficient and timely help and support during the editing process of these proceedings.

Medford, MA, USA Eric Todd Quinto
Akron, OH, USA Nathan Ida
Beijing, China Ming Jiang
Saarbrücken, Germany Alfred K. Louis

Contents

Fuzzy Rough Clustering Analysis Algorithm Based on Attribute Reduction 1
Hao Ouyang, Zhi Wen Wang, Zhen Jin Huang, and Wei Ping Hu

Electrical Resistance Tomography Based on Internal Electrode Excitation 15
Yutong Chen, Yan Han, and Botong Jing

Approximated Scale Space for Efficient and Accurate SIFT Key-Point Detection 31
Ying Wang, Yiguang Liu, Zhenyu Xu, Yunan Zheng, and Weijie Hong

The Application of Image Processing in UAV Reconnaissance Information Mining System 41
Qiufeng Pan, Jiaxing Wang, Hang Yu, Wenjun Zhang, and Peng Yue

An Object Detection Algorithm for UAV Reconnaissance Image Based on Deep Convolution Network 53
Xinping Guo, Xiangbin Li, Qiufeng Pan, Peng Yue, and Jiaxing Wang

Visualization of Horizontal Settling Slurry Flow Using Electrical Resistance Tomography 65
Kun Li, Mi Wang, and Yan Han

CSAR Moving Target Detection with Logarithm Background Subtraction Based on Optimal Azimuth Aperture Analysis 77
Wenjie Shen, Yun Lin, Fei Teng, Yanping Wang, and Wen Hong

Ship Detection Using X-Bragg Scattering Model Based on Compact Polarimetric SAR 87
Chenghui Cao, Xingpeng Mao, Jie Zhang, Junmin Meng, Xi Zhang, and Genwang Liu

Analysis of Sea Clutter Distribution and Evaluation of Ship Detection Performance for Sentinel-1 SAR Data 97
Yongxu Li, Xudong Lai, Jie Zhang, Junmin Meng, Genwang Liu, and Xi Zhang

Reflectance and Surface Normals from a Single Hyperspectral Image 105
Wenhao Xiang, Xudong Jin, and Yanfeng Gu

Characteristics Extraction of Behavior of Multiplayers in Video Football Game 115
Zhiwen Wang, Hao Ouyang, Canlong Zhang, Bowen Tang, Zhenghuan Hu, Xinliang Cao, Jing Feng, and Min Zha

A Differential Resonant Edge Detection Sensor 131
Nathan Ida

Microwave and Millimeter Wave Sensors for Nondestructive Testing and Evaluation 143
Nathan Ida

A Discussion on the Algorithm Design of Electrical Impedance Tomography for Biomedical Applications 163
Mingyong Zhou and Hongyu Zhu

Signal Processing for Diffuse Correlation Spectroscopy with Support Vector Regression 173
Peng Zhang, Zhiguo Gui, Hao Ling, Jiaxin Liu, Xiaojuan Zhang, Yiming Liu, Andi Li, and Yu Shang

Intensity Inhomogeneity Image Segmentation Based on Active Contours Driven by Self-Similarity 185
Xu Li, Hairong Liu, and Xiaoping Yang

Application of a Temperature Measurement System Based on CCD Sensor in Blast Furnace Tuyere Raceway 199
Xiaoman Cheng, Shusen Cheng, Dongdong Zhou, and Ruixuan Zhang

Weak Magnetic Stress Internal Detection Technology of Long Gas and Oil Pipelines 209
Bin Liu, Xiao Rui Yu, Lu Yao He, Li Jian Yang, and Yong Bin Que

Index 219

Fuzzy Rough Clustering Analysis Algorithm Based on Attribute Reduction

Hao Ouyang, Zhi Wen Wang, Zhen Jin Huang, and Wei Ping Hu

Abstract Fuzzy clustering analysis algorithm has good ability to solve fuzzy problems. Rough clustering analysis has good ability to solve these problems its prior knowledge is uncertain. But in the real world, there are many problems that not only are fuzzy but also are rough and uncertain; the paper combines the idea of these two algorithms. In order to improve correction of clustering, it imports attributes reduction algorithm to get importance of each attribute, and dynamically changes attribute weight by the importance. The new algorithm firstly computes fuzzy membership degree of every object and then estimates the object that belongs to lower approximation or upper approximation of one cluster. In the analysis process, the paper provides a new way to get the cluster centers, combining fuzzy and rough theory. From experiments of four UCI data sets, it is proved that the new algorithm is better effective.

Keywords Clustering · Fuzzy theory · Rough set theory · Attribute reduction

1 Introduction

Clustering analysis is different from classifying analysis that is an unsupervised analysis process. Before the analysis, it does not know the characteristics of all attributes and the importance of all kinds of attributes for the clustering analysis process [1]. The paper [2] introduces the information entropy theory to analyze the attributes of data objects and obtains the importance of each attribute, so as to dynamically adjust the analysis process. The paper [3] introduces the idea of C4.5 algorithm, uses the information gain for the continuous attributes and category attributes, so as to help the category analysis. Although the above papers dynamically analyze the importance of each attribute, they fail to effectively combine the

---

H. Ouyang (✉) · Z. W. Wang · Z. J. Huang · W. P. Hu
School of Computer Science and Communication Engineering, Guangxi University of Science and Technology, LiuZhou, Guangxi Province, PR China

E. T. Quinto et al. (eds.), *The Proceedings of the International Conference on Sensing and Imaging, 2018*, Lecture Notes in Electrical Engineering 606,
https://doi.org/10.1007/978-3-030-30825-4_1

fuzzy theory and the rough theory to deal with nonlinear problems. Fuzzy theory and rough set theory describe problems from the two different aspects: the fuzziness and the uncertainty. Therefore, a lot of papers combine fuzzy theory or rough set theory with clustering analysis algorithm to analyze problems. These papers [4–6] use fuzzy clustering algorithm to analyze medical image problem, social service problem, and carbonate fluid detection problem. These papers [7–9] use rough clustering algorithm to analyze fault detection in rotating machinery equipment, gene expression problems, and psychological anxiety problems of students. The paper [10] tries to combine the fuzzy theory and rough set theory and applies the algorithm to analyze hydrological engineering. In the process of combination, FCM algorithm is firstly adopted to obtain the mean points of fuzzy clustering, and then the rough set is calculated to obtain the upper and lower approximation of clusters. In the process, the two theories are not combined well, instead there are separated. In the paper [11], data object sets are preprocessed by using support vector machine, then introduces fuzzy clustering and rough set respectively, so the two steps are also independent. In the paper [12], fuzzy rough clustering is applied to the image analysis of brain MRI. In the process of analysis, the difference of membership degree is used to determine the rough boundary, but for calculating the mean point, only using the same method in the FCM algorithm. There are the same defect likely papers [10, 11].

In this paper, a new algorithm is provided. The reduction algorithm in rough set theory is used to obtain the importance of each attribute, dynamically adjust the weight values of each attribute in the iteration calculate process, so as to optimize the value of important attributes. In the clustering analysis, to combine the fuzzy and rough set theories, firstly it gets the fuzzy membership degree of each data object, and then gets fuzzy lower approximations and upper approximations of these clusters. The process effectively combines fuzzy and rough theory by calculating cluster mean points. This method more approximately describes the real world. The new algorithm presented in this paper is proved to have better analysis performance by several related experiments.

Chapter 2 introduces rough set theory and attribute reduction algorithm. This chapter introduces fuzzy clustering and rough clustering. In Chap. 4, fuzzy rough clustering based on reduction is introduced. In Chap. 5, a new algorithm is adopted to analyze multiple data sets. Chapter 6 is the conclusion and expectation.

2 Rough Set Theory and Reduction Algorithm

Rough set theory is a common method to analyze uncertain problems. In this theory, the most important concepts include lower approximation, upper approximation, bound and approximation quality.

Definition 1 for Low approximation, Upper approximation, and Bound. For an information system $\mathrm{IS} = (U, A, V, F)$, $B \subseteq A$, there exists equivalent relation

R_B, $\langle U, R_B \rangle$ is defined as approximate space. For any set X, $X \subseteq U$, the low approximation of X for $\langle U, R_B \rangle$ is defined as:

$$\underline{R_B}X = \{x \in U | [x]_B \subseteq X\} \quad (1)$$

The upper approximation of X for $\langle U, R_B \rangle$ is defined as:

$$\overline{R_B}X = \{x \in U | [x]_B \cap X \neq \phi\} \quad (2)$$

The bound of X for $\langle U, R_B \rangle$ is defined as:

$$BN_B = \overline{R_B}X - \underline{R_B}X \quad (3)$$

In formula (1), U is a finite object set for analysis, A is an attribute set for objects. V is a domain set for $A, f : U \times A \to V$ is an information function. By definition, $\underline{R_B}X$ is looked as the set composed by objects that certainly belong to X, $\overline{R_B}X$ is looked as the set composed by objects that probably belong to X, BN_B is looked as the set composed by objects that probably belong or not belong to X.

Definition 2, Approximate quality. $r_B(X)$ is the rough degree of set X for B, it is defined as:

$$r_B(X) = | \underline{R_B}X | / | X | \quad (4)$$

In formula (4), $| \bullet |$ is the number of elements contained in the set. If $r_B(X) = 1$, it means the bound does not exist, $BN_B = \Phi$. If $r_B(X) < 1$, it means that X is rough referring to B.

In the analysis of actual data, there is a lot of redundant information, and the importance of each attribute is different. In order to deal with problems effectively, the attribute reduction of rough set theory can be used to find the importance of attributes and eliminate the irrelevant or redundant attributes, so as to improve the correctness of clustering.

Definition 3, Attribute importance. For the decision table DT = $(U, C \cup D)$, if $B \subseteq C$ and $a \in C - B$, $\text{sig}_r(a, B, D)$ is the importance of attribute a referring to attribute set B, $\text{sig}_r(a, B, D)$ is defined as:

$$\text{sig}_r\,(a, B, D) = r_{B \cup \{a\}}(D) - r_B(D) \quad (5)$$

In formula (5), $r_B(D)$ is the dependent degree of decision attribute D referring to conditional attribute.

According to formula (5), $\text{sig}_r(a, B, D)$can be seen as the analysis contribution of attribute a in attribute set B. If the figure of $\text{sig}_r(a, B, D)$ is bigger, the attribute a is more important for decision. According to this reason, we can get an attribute reduction algorithm. The reduction algorithm starts from an empty attribute set, and then iteratively selects the attribute with the maximum importance. In order to avoid the influence of noise, the algorithm imports a stop threshold ε.

The reduction algorithm is described as follows:

Input: decision table $DT = (U, C \cup D)$, stop threshold ε
Output: set B that is a relative reduction of set C for set D

Steps:

Calculate the reliance of decision attributes on conditional attributes, $r_C(D)$;
$B \leftarrow \phi$;
while $B \subset C$ do
begin
for every $a \in C - B$ do
Calculate $\text{sig}_r(a, B, D)$;
Select the attribute a having biggest $\text{sig}_r(a, B, D)$, if many attributes meet this condition, then select the attribute a that has least number of combinations with attribute set B;
$B \leftarrow B \cup \{a\}$;
if $r_C(D) - r_B(D) \leq \varepsilon$
break;
end if
end for
for every $a \in B$ do
if $r_C(D) - r_{B-\{a\}}(D) \leq \varepsilon$
$B \leftarrow B - \{a\}$;
End if
output B;
end for
end while

3 Fuzzy Clustering and Rough Clustering

Fuzzy Clustering

Fuzzy c-mean clustering (FCM) is using fuzzy theory into clustering analysis process. In this algorithm, membership degree is used to indicate the possibility of object belonging to a certain cluster. The membership degree u_{ik} is described as how much degree of data object x_k belonged to the No. i cluster, $u_{ik} \in [0, 1]$, $\sum_{i=1}^{C} u_{ik} = 1$.

The objective function of FCM algorithm is:

$$J_m(U, V) = \sum_{i=1}^{C} \sum_{k=1}^{n} (u_{ik})^m d_{ik}^2 \tag{6}$$

In formula (6), d_{ik} is described as the distance between the object x_k and the mean v_i, it is defined as follows:

$$d_{ik}^2 = \|x_k - v_i\|^2 \tag{7}$$

In formula (6), m is the fuzzy factor, $m \in [1, \infty]$. Usually, $m = 2$. Through many iterative calculations, the mean point is obtained by the following formula:

$$\underset{\substack{1 \le i \le c \\ 1 \le k \le n}}{\forall} \quad u_{ik} = \begin{cases} (d_{ik})^{2/(1-m)} \Big/ \left[\sum\limits_{j=1}^{c} (d_{jk})^{2/(1-m)} \right], & \text{if} \quad v_i \ne x_k; \\ 1, & \text{if} \quad v_i = x_k; \\ 0, & \text{if} \quad v_l = x_k, l \ne i; \end{cases} \tag{8}$$

and

$$\underset{1 \le i \le c}{\forall} \; v_i = \sum_{k=1}^{n} (u_{ik})^m x_k \Big/ \sum_{k=1}^{n} (u_{ik})^m \tag{9}$$

Rough Clustering

Rough c-mean clustering (RCM) is using rough theory into clustering analysis process. In the algorithm, it is agreed that an object can only belong to the lower approximation of one cluster mostly. When calculating the distance between the data object and each mean point, if the difference between these distances is less than a certain threshold value, the object will be divided into the upper approximation of each corresponding cluster. Otherwise, it is divided into the lower approximation of nearest cluster.

According to the definition of lower approximation, upper approximation, and bound in formula (1)–(3), the new definition of mean point in RCM is obtained as follows:

$$\underset{1 \le i \le c}{\forall} \; v_i = \begin{cases} w_l \dfrac{\sum\limits_{x_k \in \underline{C_i}} x_k}{|\underline{C_i}|} + w_{bn} \dfrac{\sum\limits_{x_k \in C_i^{BN}} x_k}{|C_i^{BN}|}, & \text{if} \quad \underline{C_i} \ne \phi \wedge C_i^{BN} \ne \phi; \\ \dfrac{\sum\limits_{x_k \in \underline{C_i}} x_k}{|\underline{C_i}|}, & \text{if} \quad \underline{C_i} \ne \phi \wedge C_i^{BN} = \phi; \\ \dfrac{\sum\limits_{x_k \in C_i^{BN}} x_k}{|C_i^{BN}|}, & \text{if} \quad \underline{C_i} = \phi \wedge C_i^{BN} \ne \phi; \end{cases} \tag{10}$$

In the above formula, w_l is the weight of lower approximation, w_{bn} is the weight of bound. Then $w_l + w_{bn} = 1$. If w_{bn} is larger, the prior knowledge is less complete, and the information set is rougher.

4 Fuzzy Rough Clustering Based on Attribute Reduction

Fuzzy Rough Clustering

In practical application scenarios, there are many problems often lacking prior knowledge that are fuzzy description. In this case, if only fuzzy clustering or rough clustering is used, it will inevitably lead to one-sidedness of the analysis process. In this paper, the fuzzy theory and rough set theory are applied to clustering analysis, and a new fuzzy clustering algorithm is designed. In the algorithm, two equivalent classes can be obtained for the set U: they are upper approximation and non-upper approximation; upper approximation can be divided continually into lower approximation and bound; this process is defined in formula (11). However, the definition of mean point is combined with the fuzzy theory and the rough theory. In the analysis, it is believed that the fuzzy degree of data objects belong to the lower approximation, boundary or non-upper approximation of each cluster is different, and the calculation method of fuzzy membership degree is shown in formula (12).

$$\begin{aligned}\underset{1\le i\le c}{U} &= \left\{x_k | x_k \in \overline{C_i}(X)\right\} \cup \left\{x_k | x_k \notin \overline{C_i}(X)\right\} \\ &= \left\{x_k | x_k \in \underline{C_i}\right\} \cup \left\{x_k | x_k \in C_i^{BN}\right\} \cup \left\{x_k | x_k \notin \overline{C_i}(X)\right\}\end{aligned} \tag{11}$$

$$\underset{1\le i\le c}{\forall} v_i = \begin{cases} \left(w_l \dfrac{\sum\limits_{x_k\in \underline{C_i}} (u_{ik})^m x_k}{|\underline{C_i}|} + w_{bn} \dfrac{\sum\limits_{x_k\in C_i^{BN}} (u_{ik})^m x_k}{|C_i^{BN}|} + \sum\limits_{x_k\notin \overline{C_i}} (u_{ik})^m x_k \right) \Big/ \sum\limits_{k=1}^{n} (u_{ik})^m, \\ \text{if} \quad \underline{C_i} \ne \phi \wedge C_i^{BN} \ne \phi; \\ \left(\dfrac{\sum\limits_{x_k\in \underline{C_i}} (u_{ik})^m x_k}{|\underline{C_i}|} + \sum\limits_{x_k\notin \overline{C_i}} (u_{ik})^m x_k \right) \Big/ \sum\limits_{k=1}^{n} (u_{ik})^m, \\ \text{if} \quad \underline{C_i} \ne \phi \wedge C_i^{BN} = \phi; \\ \left(\dfrac{\sum\limits_{x_k\in C_i^{BN}} (u_{ik})^m x_k}{|C_i^{BN}|} + \sum\limits_{x_k\notin \overline{C_i}} (u_{ik})^m x_k \right) \Big/ \sum\limits_{k=1}^{n} (u_{ik})^m, \\ \text{if} \quad \underline{C_i} = \phi \wedge C_i^{BN} \ne \phi; \end{cases} \tag{12}$$

In the formula (12), u_{ik} is got in formula (8).

When analyzing, the fuzzy rough clustering algorithm needs to comply with the following three rules:

(a) A data object can only belong to the lower approximation of a cluster.
(b) If the data object belongs to the lower approximation of a cluster, it must also belong to the upper approximation of the cluster.
(c) When the difference between the maximum fuzzy membership of a cluster (it is u_{Max}) and the fuzzy membership of other cluster (it is u_i) is less than a certain threshold value, the data object can belong to the upper approximation of two or more clusters.

In this paper, the threshold in point C above is called the fuzzy rough threshold, which is denoted as δ.

Fuzzy Rough Clustering Based on Attribute Reduction

In the paper, a fuzzy rough clustering based on attribute reduction (FRCMbR) is proposed. In the calculation process, the algorithm obtains the importance of each attribute by the reduction algorithm. Then it gives different weight value to each attribute according to the importance. At last, it makes corresponding adjustments for the distance between data object and cluster by different weight value. The new difference is calculated as shown in formula (13).

$$d_{ik}^2 = \sum_{p=1}^{\text{cond}} r_p \left(x_{kp} - v_{ip}\right)^2 \tag{13}$$

where r_p is the weight of the No. p attribute of the object, cond is the count of the attributes, and

$$1 = \sum_{p=1}^{\text{cond}} r_p \tag{14}$$

Fuzzy rough theory is introduced in clustering analysis.

The fuzzy rough clustering algorithm is described as follows:

Input: Set U has n data objects; number of clusters (c); lower approximate weight (w_l); bound weight (w_{bn}); fuzzy rough threshold (δ); algorithm stop threshold (ε).

Output: c clusters

Steps:

Initialize the weight of each attribute of a data object: $r_p \leftarrow 1/n$;

Randomly generate the c mean points of each cluster,
The number of iterations: $j \leftarrow 0$;
do
Running the reduction algorithm to get the weight of each attribute;
Using formula (13) calculate the difference between data objects and each cluster;
Using formula (8) get the membership degree of each object (u_{ik}) in the No. j iteration calculation;
if $|u_{ik} - u_{mk}| \leq \delta$ && $i \neq m$
$\overline{C_i} \leftarrow \overline{C_i} \cup \{x_k\}$;
$\overline{C_m} \leftarrow \overline{C_m} \cup \{x_k\}$;
else
$\underline{C_i} \leftarrow \underline{C_i} \cup \{x_k\}$
end if-else
Using formula (12) get the mean points of each cluster again;
The number of iterations: $j \leftarrow j + 1$;
while $\left(\max\left\{|u_{ik}^{(j)} - u_{ik}^{(j-1)}|\right\} > \varepsilon\right)$;
end do-while;

5 Experiments of FRCMbR Algorithm

Experimental Setting

The computer configuration for testing the algorithm: CPU is 2.93 GHz dual–core, and the memory is 4 GB. Software configuration: Window 7, Matlab2007. The experimental UCI data sets are Iris, Nursery, Ionosophere, and Isolet5, respectively. The characteristics of the data sets are shown in Table 1.

The definition of accuracy in the experiment is as follows:

$$\text{accurate}(U) = \frac{\sum_{k=1}^{c} \text{correct}_i}{|U|} \tag{15}$$

In formula (15), $|U|$ is the number of data sets, and correct_i is the number of data objects correctly divided into corresponding clusters.

Table 1 Characterization of four UCI sets

Data set	Number of samples	Attribute dimension	Number of clusters	Instructions
Iris	150	4	3	Data on Iris flowers
Nursery	12,960	8	3	Nursery data
Ionosophere	351	34	2	Radar acquisition data
Isolet5	1559	617	26	Alphabetic predictive data

Table 2 Comparison of correction for Iris set

Algorithm	Number of samples	Correct samples	Wrong samples	Accuracy (%)
K-Means	150	105	45	70.0
FCM	150	130	20	82.7
RCM	150	129	21	86.0
FRCM	150	132	18	88.0
FRCMbR	150	139	11	92.7

Table 3 Comparison of correction for Nursery set

Algorithm	Number of samples	Correct samples	Wrong samples	Accuracy (%)
K-Means	12,960	8451	4509	65.2
FCM	12,960	8935	4025	68.9
RCM	12,960	9051	3909	69.8
FRCM	12,960	10,455	4170	80.7
FRCMbR	12,960	10,561	2399	81.4

Table 4 Comparison of correction for Ionosophere set

Algorithm	Number of samples	Correct samples	Wrong samples	Accuracy (%)
K-Means	351	207	144	59.0
FCM	351	234	117	66.7
RCM	351	241	110	68.7
FRCM	351	249	102	70.9
FRCMbR	351	292	59	83.2

Table 5 Comparison of correction for Isolet5 set

Algorithm	Number of samples	Correct samples	Wrong samples	Accuracy (%)
K-Means	1559	894	665	57.3
FCM	1559	1032	527	61.2
RCM	1559	1025	534	65.7
FRCM	1559	1098	461	70.4
FRCMbR	1559	1286	273	82.4

Comparison of the Accuracy of Each Algorithm

Tables 2, 3, 4, and 5 shows the accuracy of different algorithm after analysis of the above four data sets. For the four experiments of FRCMbR algorithm, the stooping threshold $\varepsilon = 0.00001$, the lower approximate weight $w_l = 0.9$, and the bound weight $w_{bn} = 0.1$.

According to the analysis results in Tables 2, 3, 4, and 5, the traditional FRCM algorithm combines the fuzzy theory with the rough theory, which makes the analysis more robust. Compared with FCM and RCM, it is more capable of analyzing nonlinear problems, so its analysis results are more accurate. The FRCMbR algorithm proposed in this paper is superior to the other four algorithms because it not only recombines the fuzzy and rough theory, but also conforms to the concept of fuzzy roughness and is close to the real world. In the process of analysis, the weights of each attribute are dynamically adjusted according to the analysis results, so as to optimize the analysis results. From the perspective of Iris and Nursery data sets, if the attribute dimension of the data set is small, the FRCMbR algorithm is superior to the FRCM algorithm, but its advantages are not particularly obvious. When the attribute dimension increases, the FRCMbR algorithm's analytical capability gradually shows that it is superior to the FRCM algorithm in a word, FRCMbR algorithm has two advantages:

(a) FRCMbR algorithm combines the characteristics of fuzzy theory and rough theory, making it better able to deal with nonlinear problems, especially when the background knowledge is not clear, fuzzy, and incomplete, it has more advantages.
(b) FRCMbR algorithm introduces reduction algorithm, which can dynamically adjust the weight of attributes according to the actual importance in the analysis, so as to strengthen the major attributes on final analysis results.

Parameter Experiments of FRCMbR Algorithm

The important parameters to be adjusted in FRCMbR analysis include: lower approximate weight w_l, bound weight w_{bn}, fuzzy rough threshold δ, algorithm stop threshold ε. Where the algorithm stop threshold ε generally selects a minimum value approaching 0. The approximate weight w_l, bound weight w_{bn}, and fuzzy rough threshold δ are directly related to rough calculation. Therefore, this paper conducts experiments on these three parameters to further verify the validity of FRCMbR algorithm. Among the three parameters, w_l and w_{bn} are correlated, $w_l + w_{bn} = 1$, so when analyzing these two parameters, only one of them needs to be set, and the other one can be determined.

In order to test the w_l and w_{bn}, it sets $\delta = 0.1$. The algorithm deals with data set Iris, Nursery, Ionosphere, and Isolet5, respectively, 20 times each for each data set, and then average the results. These test results are shown in Fig. 1.

As can be seen in Fig. 1, when it deals with Iris data set, it sets $w_{bn} = 0.1$, the analysis accuracy 92.7% is the highest; when it deals with Nursery data set, it sets $w_{bn} = 0.15$, the analysis accuracy 82.7% is the highest; when it deals with Ionosphere data set, it sets $w_{bn} = 0.15$, the analysis accuracy 85.9% is the highest; when it deals with Isolet5data set, it sets $w_{bn} = 0.2$, the analysis accuracy 88.4% is the highest. Conclusion can be drawn from the above analysis:

(a) When w_{bn} is within the $[0.1, 0.2]$ range, the accuracy is highest. This is because only a small part of the data object that clusters in the whole problem domain is uncertain, but the overall information is uncertain. The impact of these uncertain data objects on the overall analysis results can be small.

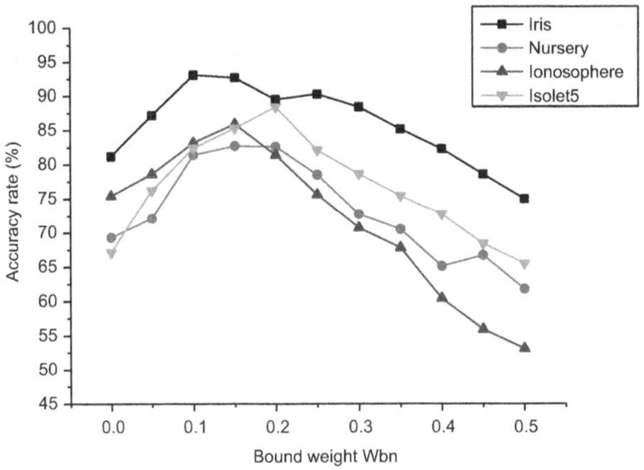

Fig. 1 Influence of boundary weight for correction

Fig. 2 Influence of fuzzy rough threshold for correction

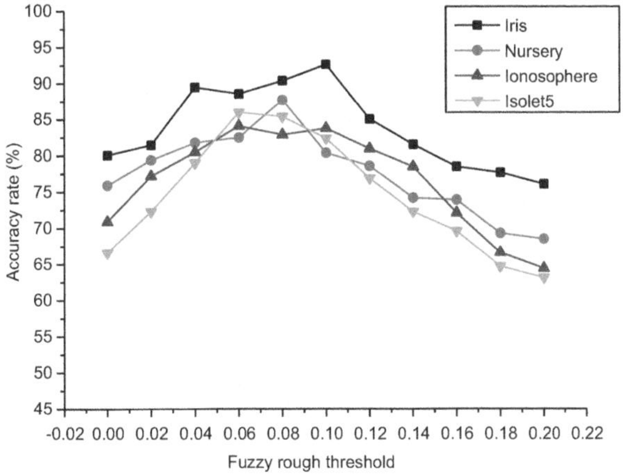

(b) At that time $w_{bn} = 0$, FRCMbR algorithm is focused only on the data objects determined by prior knowledge, completely ignoring the bound objects in rough set. And the algorithm is reduced to a fuzzy clustering, that only dealt with lower approximation of each cluster.
(c) At that time $w_{bn} > 0.25$, the analytical capability of FRCMbR algorithm is gradually declined. This is because, in the process, the algorithm will only pay more attention to these objects whose prior knowledge is uncertain, but easing to ignore the objects of the low approximation.

In the fuzzy rough threshold δ test, this paper sets the boundary weight $w_{bn} = 0.1$, and the algorithm tests the four data sets, respectively. The test results are shown in Fig. 2.

As can be seen from Fig. 2 when $\delta = 0.1$ in the test for Iris data set, the highest accuracy rate is 92.7%; when $\delta = 0.08$ in the test for Nursery data set, the highest accuracy rate is 87.7%; when $\delta = 0.06$ in the test for Ionosophere data set, the highest accuracy rate is 84.1%; when $\delta = 0.06$ in the test for Isolet5 data set, the highest accuracy rate is 86.0%. The conclusion is drawn from the above analysis:

(a) When the δ value range within the $[0.06, 0.1]$, the accuracy is the highest. This is because most of the data is not fuzzy and uncertainty, only the data of bound that is fuzzy.
(b) Comparing with the bound weight w_{bn}, the fuzzy rough threshold δ is a smaller values The reason is that the δ is a difference of fuzzy membership degrees for the data object belonging to different clusters. (e.g., $|u_{ik} - u_{ij}|$ is the difference of fuzzy membership degree for the No. i data object belonging to the No. k or the No. j cluster). The fuzzy membership degree value is small and it is in [0, 1], so the δ is more smaller.
(c) At that time $\delta = 0$, the FRCMbR algorithm is no longer capable of rough analysis, and the algorithm is degraded to fuzzy clustering algorithm.
(d) At that time $\delta > 0.1$, the analysis capability of FRCMbR algorithm is also declining. This is because as δ increases, the algorithm divides more data objects into the bound, believing that the whole analysis object is more rough, which is inconsistent with the actual situation.

6 Conclusion

Fuzzy theory and rough set theory helps to understand the real world from different angles. When combined with K-Means algorithm, the two theories can be selected according to different scenarios. Fuzzy K-Means is better at the ambiguous and paradoxical problems, while rough K-Means is better at the problems of incomplete prior knowledge or incomplete prerequisite knowledge. The FRCMbR algorithm proposed in this paper combines fuzzy theory and rough set theory and deals with problems from two aspects at the same time. FRCMbR algorithm also introduces the reduction algorithm in rough set theory, in order to obtain the important degree of each attribute, then, dynamically adjust their weight, and further optimize the important attribute. The FRCMbR algorithm is verified through several experiments on UCI data sets. However, in the process of algorithm calculation, the parameter setting will affect the final analysis results. Therefore, it is necessary to further improve the algorithm in future research to reduce the influence of experience factors on the algorithm.

Acknowledgments Young Teachers' Basic Ability Improving Project of Guangxi Education Hall under Grant No. 2018KY0321; The National Natural Science Foundation of China (61462008, 61751213, 61866004); The Key projects of Guangxi Natural Science Foundation (2018GXNSFDA294001,2018GXNSFDA281009); The Natural Science Foundation of Guangxi (2017GXNSFAA198365); 2015 Innovation Team Project of Guangxi University of Science and Technology (gxkjdx201504); Scientific Research and Technology Development Project of Liuzhou (2016C050205); Guangxi Colleges and Universities Key Laboratory of Intelligent Processing of

Computer Images and Graphics under Grant No. GIIP201508; Young Teachers' Basic Ability Improving Project of Guangxi Education Hall under Grant No. KY2016YB252; Natural Science Foundation of Guangxi University of Science and Technology under Grant No. 174523.

References

1. Han, J.-w., Kamber, M., & Pei, J. (2012). *Data mining: Concept and techniques* (3rd ed.). Beijing: China Machine Press.
2. Ouyang, H., Dai, X.-S., Wang, Z.-w., & Wang, M. (2015). Rough K-prototypes clustering algorithm based on entropy. *Computer Engineering and Design, 36*(5), 1239–1243.
3. Ouyang, H., Wang, Z.-w., Dai, X.-S., & Liu, Z.-q. (2015). A fuzzy K-prototypes clustering algorithm based on information gain. *Computer Engineering & Science, 37*(5), 1009–1014.
4. Kannan, S. R., Devi, R., Ramathilagam, S., & Takezawa, K. (2013). Effective FCM noise clustering algorithms in medical images. *Computers in Biology and Medicine, 43*(2), 73–83.
5. Ghodousi, M., Alesheikh, A. A., & Saeidian, B. (2016). Analyzing public participant data to evaluate citizen satisfaction and to prioritize their needs via K-means, FCM and ICA. *Cities, 55*(6), 70–81.
6. Liu, L., Sun, S. Z., Yu, H., Yue, X., & Zhang, D. (2016). A modified Fuzzy C-Means (FCM) clustering algorithm and its application on carbonate fluid identification. *Journal of Applied Geophysics, 129*(6), 28–35.
7. Pacheco, F., Cerrada, M., Sánchez, R.-V., Cabrera, D., Li, C., & de Oliveira, J. V. (2017). Attribute clustering using rough set theory for feature selection in fault severity classification of rotating machinery. *Expert Systems with Applications, 71*(4), 69–86.
8. Nayak, R. K., Mishra, D., Shaw, K., & Mishra, S. (2012). Rough set based attribute clustering for sample classification of gene expression data. *Procedia Engineering, 38*, 1788–1792.
9. Yanto, I. T. R., Vitasari, P., Herawan, T., & Deris, M. M. (2012). Applying variable precision rough set model for clustering student suffering study's anxiety. *Expert Systems with Applications, 39*(1), 452–459.
10. Wang, H., Feng, Q., Lin, X., & Zeng, W. (2009). Development and application of ergodicity model with FRCM and FLAR for hydrological process. *Science China: Technological Scinences, 52*(2), 379–386.
11. Saltos, R., & Weber, R. (2016). A rough–fuzzy approach for support vector clustering. *Information Sciences, 339*(4), 353–368.
12. Dubey, Y. K., Mushrif, M. M., & Mitra, K. (2016). Segmentation of brain MR images using rough set based intuitionistic fuzzy clustering. *Biocybernetics and Biomedical Engineering, 36*(2), 413–426.

Electrical Resistance Tomography Based on Internal Electrode Excitation

Yutong Chen, Yan Han, and Botong Jing

Abstract Electrical resistance tomography (ERT) reconstructs the resistance/conductivity distribution in the sensitive field by measuring the boundary voltage of the field. In the traditional ERT adjacent excitation mode, the sensitivity coefficients amplitude of the central region of the field is low, which leads to the poor imaging resolution of the central region. In this chapter, the influence of the distribution of the sensitivity coefficients amplitude of ERT system on image reconstruction is analyzed. A new excitation mode is proposed, in which the inner electrode is added to the center of the field to improve the sensitivity amplitude of the central region and thus improve the quality of image reconstruction. Aiming at the problem of central artifact caused by internal electrode excitation, the mixed excitation mode and its corresponding image reconstruction algorithm are used to reduce it. The simulation results show that the new excitation mode and algorithm can effectively improve the resolution of field center image and the quality of reconstructed image.

Keywords Electrical resistance tomography · Sensitivity matrix · Internal electrode excitation · Mixed excitation · Reconstruction algorithm

---

Y. Chen (✉)
Shanxi Key Laboratory of Signal Capturing and Processing, North University of China, Taiyuan, People's Republic of China

Department of Electrical Engineering and Automation, Shanxi Polytechnic College, Taiyuan, People's Republic of China

Y. Han · B. Jing
Department of Electrical Engineering and Automation, Shanxi Polytechnic College, Taiyuan, People's Republic of China

E. T. Quinto et al. (eds.), *The Proceedings of the International Conference on Sensing and Imaging, 2018*, Lecture Notes in Electrical Engineering 606,
https://doi.org/10.1007/978-3-030-30825-4_2

1 Introduction

Electrical resistance tomography (ERT) [1] reconstructs the conductivity distribution in the field by measuring the boundary voltage of the field. ERT system has been widely used in industrial surveying [2–4], geological exploration [5–7], and other fields because of its simple equipment and fast imaging speed. However, due to the low resolution of ERT reconstruction image, it is very limited in application. In recent years, in order to improve the image resolution of ERT, experts have done a lot of research on its excitation measurement mode. At present, the adjacent measurement with adjacent excitation mode [8] is commonly used in measurement. In addition, Liu proposed a single-electrode excitation multi-electrode measurement model [9], and Ren proposed a new sensor with 8 internal electrodes and 16 external electrodes [10].

In this chapter, the distribution characteristics of sensitivity coefficient of ERT system are analyzed. To solve the problem of low sensitivity coefficient in central field, a measurement mode based on internal electrode excitation is proposed. We propose adding an inner electrode in the center of the field as the excitation current outflow terminal. This method can increase the magnitude of sensitivity coefficient in the central field and improve the quality of image reconstruction. To solve the central artifact problem caused by internal electrode excitation, a new algorithm is proposed. In the new algorithm, the adjacent excitation and inner electrode excitation are used to measure the object, and then the image reconstruction algorithm is improved to reduce artifacts. The simulation results show that the new excitation mode and algorithm can effectively improve the resolution of the reconstructed image in the central field and improve the quality of the reconstructed image.

2 The Effect of ERT Sensitivity Coefficient Amplitude on Image Reconstruction

Principle of ERT

For ERT systems, adjacent measurement mode with adjacent excitation is commonly used as shown in Fig. 1 [8]. The current is injected into the measured area from one pair of adjacent electrodes, and the voltage between the other adjacent electrodes is measured at the same time. The measurement is then stimulated until all the pairs of electrodes are used as one-time excitation electrodes. One hundred and four independent measuring voltages can be obtained in 16-electrode measuring system [11].

Based on the principle of sensitivity coefficient, when the conductivity distribution in the measured area changes very little, the linear model of current excitation and voltage measurement mode (Fig. 1) can be expressed as [12].

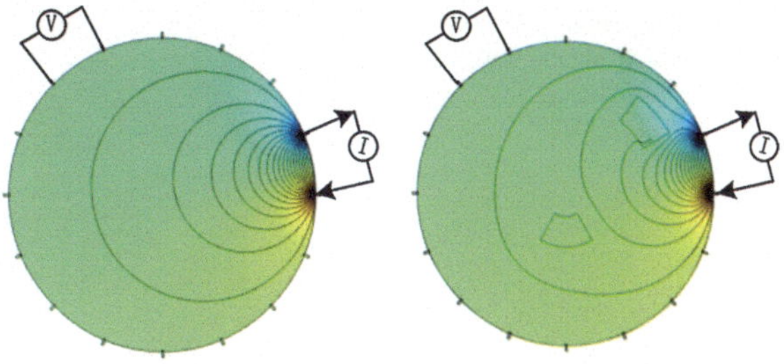

Fig. 1 16-electrodes ERT model. (**a**) Uniform conductivity. (**b**) Perturbed conductivity

$$\mathbf{S} \cdot \Delta\sigma = \Delta\mathbf{Z} \tag{1}$$

where $\Delta\mathbf{Z}$ is the changed vector of measured boundary voltage, $\Delta\sigma$ is the changed conductivity vector, and $\mathbf{S}$ is the calculated sensitivity matrix. After the conductivity change vector is obtained, the conductivity distribution vector can be obtained from Eq. (2).

$$\sigma = Q\left[\sigma_0 + \Delta\sigma\right] \tag{2}$$

where σ_0 is the initial conductivity distribution in the field, and $Q[]$ is a projection operator [13], which limits the conductivity distribution in the field between 0 and 1.

The Effect of Sensitivity Coefficient Amplitude Distribution on Image Reconstruction

$\mathbf{S} = [\mathbf{S}_{ij}]$ is the calculated sensitivity matrix, where $\mathbf{S}_{ij}$ is the sensitivity coefficient of the j-th pixel relative to the i-th measurement. Assuming that the m-th pair of electrodes is used as the exciting pair and the n-th pair of electrodes is used as the measurement electrode pair in the i-th measurement, then $\mathbf{S}_{ij}$ can be calculated by [12]

$$\mathbf{S}_{ij} = -\int \frac{\nabla\varphi_m}{\mathbf{I}_n} \cdot \frac{\nabla\psi_n}{\mathbf{I}_n} dxdy \tag{3}$$

where φ_m and ψ_n represent the potential distribution in the field when the current is $\mathbf{I}_m$ and $\mathbf{I}_n$, respectively, when the m-th and n-th pairs of electrodes are used as exciting electrodes.

As the amount of boundary measurement data of ERT system is far less than the number of finite element to be solved, image reconstruction is seriously undetermined [14] and cannot be solved directly by inverse matrix. So the ERT reconstruction often uses iterative method to solve the approximate solution of image, such as Landweber algorithm [15]

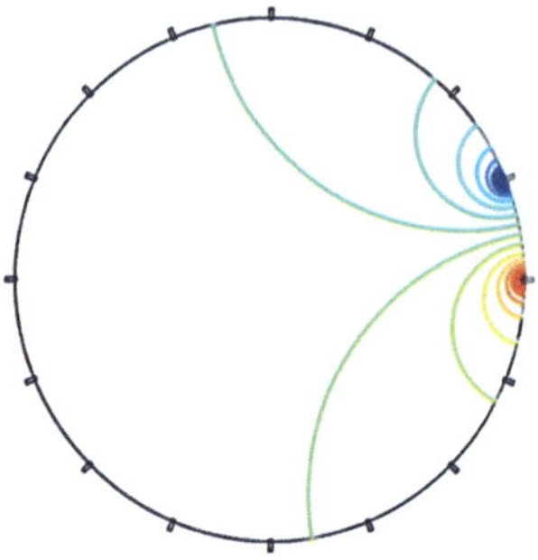

Fig. 2 Equipotential lines of adjacent electrode excitation

$$\begin{cases} \Delta\sigma_{k+1} = \Delta\sigma_k + \alpha \cdot \mathbf{S}^{\mathrm{T}} (\Delta\mathbf{Z} - \mathbf{S} \cdot \Delta\sigma_k) \\ \Delta\sigma_1 = \mathbf{S}^{\mathrm{T}} \Delta\mathbf{Z} \end{cases} \tag{4}$$

where $\Delta\sigma_k$ is the conductivity change vector obtained by step k iteration and the constant α is known as the gain factor and is used to control the convergence rate.

$\Delta\sigma_{2,\,l}$ represents the variation of the l-th finite element obtained by the second iteration, and $\mathbf{S}_l$ represents the l-th column of the sensitivity matrix $\mathbf{S}$, which consists of 104 sensitivity coefficients of the l-th finite element, from Eq. (4).

$$\Delta\sigma_{2,l} = \mathbf{S}_l^{\mathrm{T}} \cdot \left(\mathbf{I} + \alpha \cdot \mathbf{I} - \alpha \cdot \mathbf{S} \cdot \mathbf{S}^{\mathrm{T}}\right) \cdot \Delta\mathbf{Z} \tag{5}$$

can be obtained. The larger the amplitude of each element in $\mathbf{S}_l$, the larger the amplitude of $\Delta\sigma_{2,\,l}$. The principle of subsequent iteration is similar. In addition, Eq. (3) shows that the magnitude of $|\mathbf{S}_{ij}|$ which is positively correlated with the magnitude of the potential gradients $|\nabla\varphi_m|$ and $|\nabla\psi_n|$ of the finite element method. Therefore, it can be seen that the magnitudes of $\nabla\varphi_m$ and $\nabla\psi_n$ will affect the effect of image reconstruction.

Image Reconstruction Analysis of Adjacent Excitation Modes

When the adjacent excitation mode is used, the equipotential line in the field is shown in Fig. 2.

From the above analysis, it can be seen that the closer the excitation electrode is, the more dense the electric field equipotential lines are, the bigger the $|\nabla\varphi_m|$ and $|\nabla\psi_n|$ are, and the bigger the $|\mathbf{S}_{ij}|$ is. Correspondingly, within the field far from the electrode, the more sparse equipotential lines are, the smaller the $|\nabla\varphi_m|$ and $|\nabla\psi_n|$ are, and the smaller the $|\mathbf{S}_{ij}|$ is. The field is divided into 16 circles and the number of finite elements is 1536. The finite element division is shown in Fig. 3a. The finite element is numbered in order from the center to the outside. The smaller the finite element number is, the closer the finite element is to the central field. The amplitude

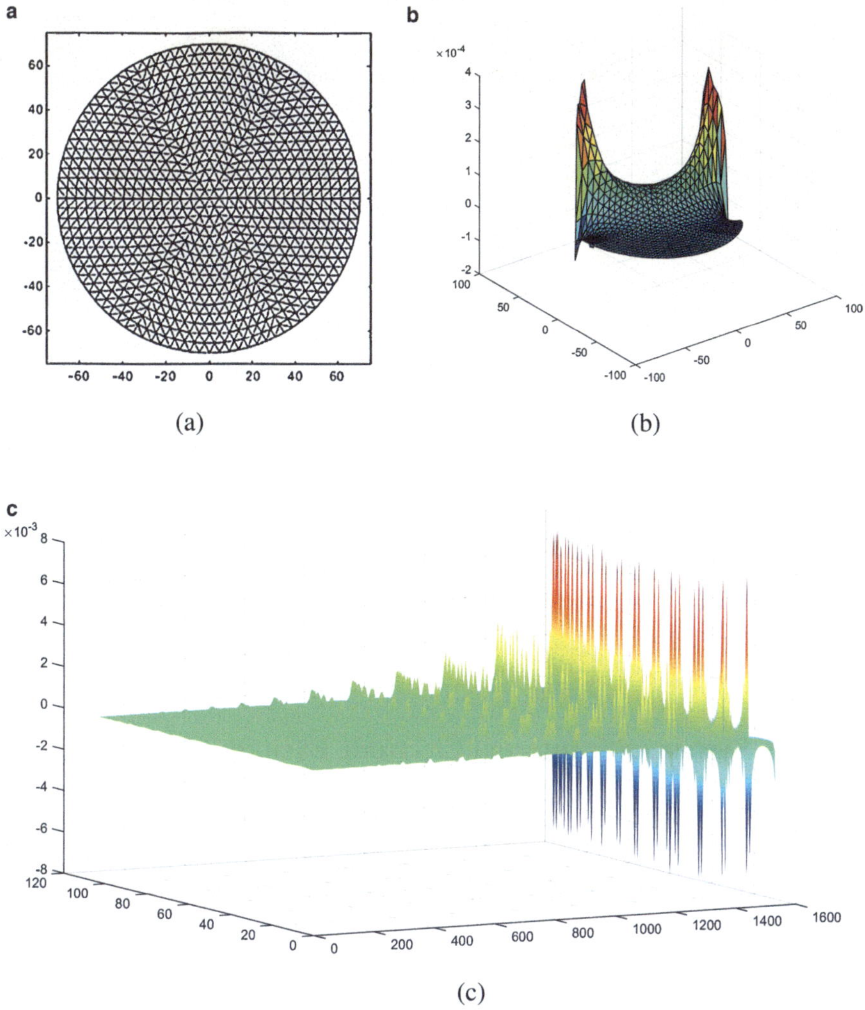

Fig. 3 Sensitivity coefficient distribution with adjacent excitation mode. (**a**) Mesh grid. (**b**) Amplitude distribution of sensitivity coefficient in once measurement. (**c**) Amplitude of sensitivity coefficient of adjacent measurement under adjacent excitation

distribution of sensitivity coefficient in once measurement is shown in Fig. 3b. After 104 measurements, the sensitivity coefficient amplitude distribution of 1536 finite elements is obtained as shown in Fig. 3c.

Taking the model of Fig. 4a as an example, the effect of adjacent excitation modes on the reconstruction of discrete phases at different positions in the field is analyzed. The normalized discrete phase conductivity is 0 and the continuous phase conductivity is 1. The reconstructed image is shown in Fig. 4b. The contour of reconstructed image is clear because of the dense equipotential lines near the

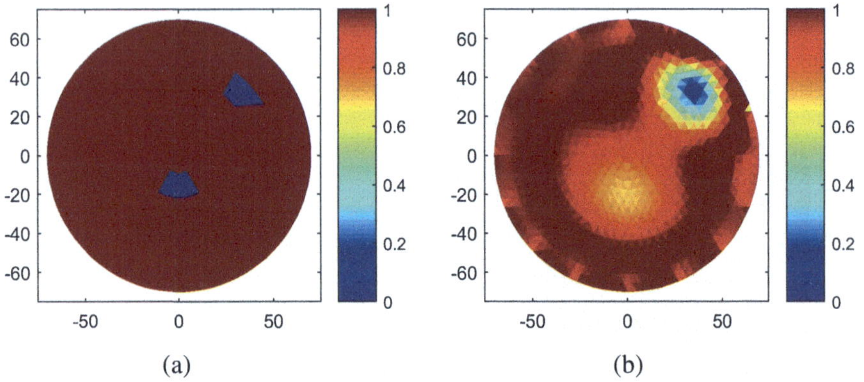

Fig. 4 Reconstructed image of adjacent excitation adjacent measurement mode. (**a**) Model D1. (**b**) Reconstructed image of model D1

electrodes and the large variation of sensitivity coefficient amplitude. However, the equipotential line near the center of the field are sparse and the sensitivity coefficient amplitude changes a little, so the reconstructed image is blurred and the contours are difficult to distinguish.

3 ERT System with Internal Electrode Excitation

Internal Electrode Excitation

In order to improve the central field, a measurement mode with internal electrode excitation is proposed in this chapter. Taking flow detection as an example, in order to enhance the sensitivity of the central field, an inner electrode can be added to the center of the field. The structure is shown in Fig. 5.

The inner electrode is used as the current outflow terminal, and the 16 electrodes around the field are used as the current input terminal to excite in turn. The boundary voltage is measured by adjacent measurement mode, as shown in Fig. 6.

Image Reconstruction of Internal Electrode Excitation Mode

The distribution of field contour is shown in Fig. 7 when the inner electrode is used for excitation. Like adjacent excitation measurement mode, the closer the excitation electrode is, the more dense the contour lines are. However, due to the addition of central electrodes, the distance between each finite element and the electrodes in the field is reduced, so the distribution of potential lines in the field is more uniform.

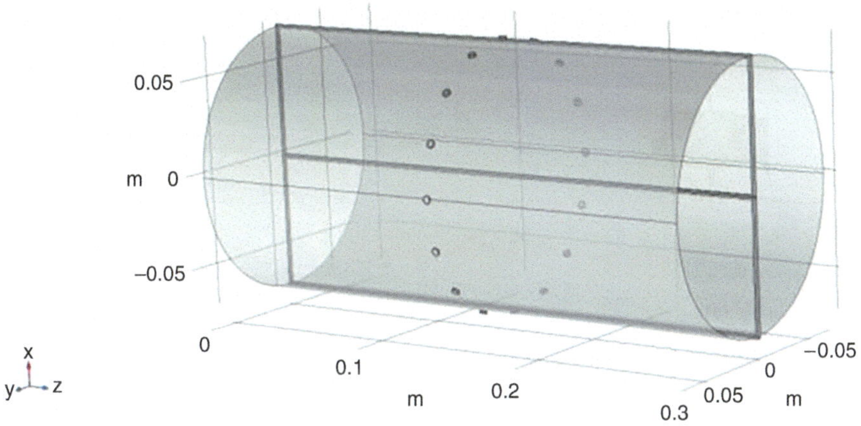

Fig. 5 ERT system with internal electrodes

Fig. 6 ERT model with electrode excitation. (**a**) Uniform conductivity. (**b**) Perturbed conductivity

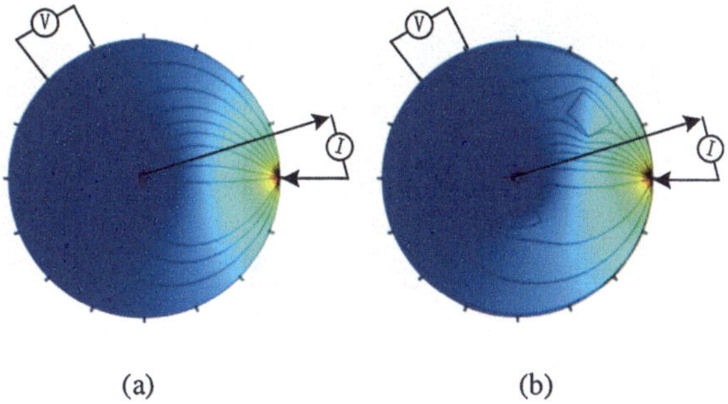

Fig. 7 Equipotential lines of internal electrode excitation

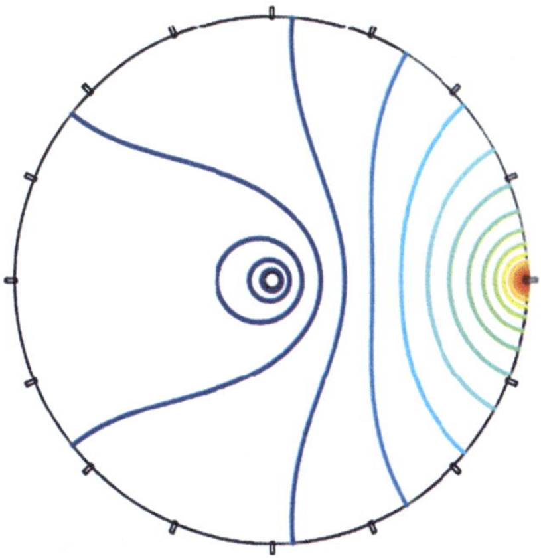

The distribution of sensitivity coefficient is calculated by Eq. (3). The distribution of sensitivity coefficient in once measurement is shown in Fig. 8a. After 224 measurements, the sensitivity coefficient amplitude distribution of 1536 finite

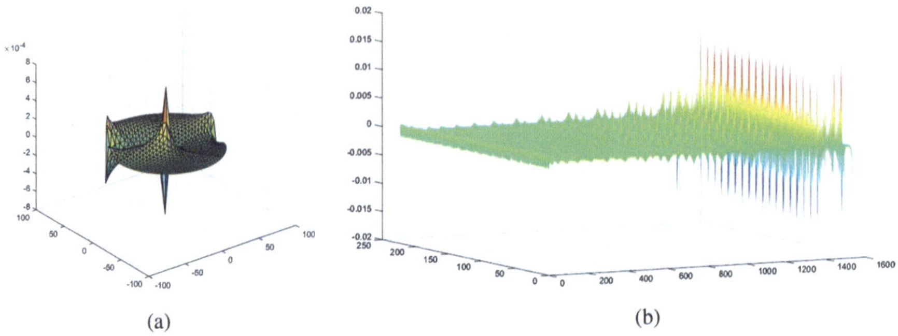

Fig. 8 Sensitivity coefficient distribution with adjacent excitation mode. (**a**) Amplitude distribution of sensitivity coefficient in once measurement. (**b**) Amplitude of sensitivity coefficient of adjacent measurement under adjacent excitation

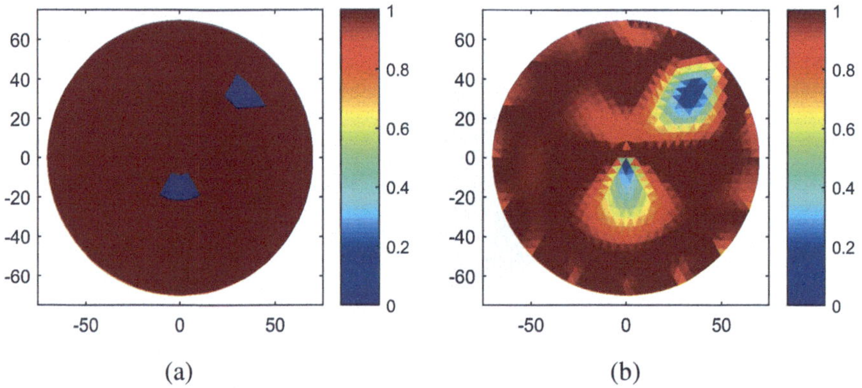

Fig. 9 Reconstructed image of inner electrode excitation adjacent measurement mode. (**a**) Model D1. (**b**) Reconstructed image of model D1

elements is obtained as shown in Fig. 8b. It can be seen that the magnitude of sensitivity coefficient near the center of the field is enhanced.

The model D1 in Fig. 4a is reconstructed by using the inner electrode excitation measurement mode with Eq. (4) as shown in Fig. 9b.

Compared with the adjacent excitation mode, the inner electrode excitation mode significantly enhances the imaging effect in the central field, and the image boundary is clear. But it will produce artifacts which tend to the inner electrode. In the two-phase flow system model, the conductivity of continuous phase is 1 and that of discrete phase is 0. Artifacts in this chapter refer to the area where the original conductivity value should be 1, and the conductivity value is less than 1 due to reconstruction error. As a result, the discrete phase distribution in the reconstructed image is enlarged.

The reconstructed conductivity near the electrode changes dramatically because the sensitivity coefficient is much higher than that of other positions in the field. Artifacts will be generated near the electrode whether the image is reconstructed by the adjacent excitation mode or the internal electrode excitation mode. Although the inner electrode excitation mode can significantly improve the image reconstruction effect in the central field, the artifacts near the inner electrode are also serious. If the discrete phase is distributed near the electrode, the image will be distorted, as shown in Fig. 9b.

Algorithms for Mixed Excitation Mode

In order to reduce the artifacts near the inner electrode, the results of conductivity distribution of adjacent excitation modes are used to modify the model.

Firstly, the internal electrode excitation mode and the adjacent excitation mode are used to reconstruct the structure, respectively. When the adjacent excitation mode is adopted, the internal electrode is in the open circuit state. The conductivity distributions of two reconstructions are σ_{Adjacent} and σ_{Internal}, respectively.

Then, in each finite element, the maximum value $\sigma_{i,\text{ Max}}$ between $\sigma_{i,\text{ Adjacent}}$ and $\sigma_{i,\text{ Internal}}$ is taken as the reconstruction distribution.

$$\sigma_{i,\text{Max}} = \text{Max}\left[\sigma_{i,\text{Adjacent}}, \sigma_{i,\text{Internal}}\right] \quad (i = 0, 1, \ldots, 1536) \tag{6}$$

At the center of the field, $\sigma_{i,\text{ Adjacent}} > \sigma_{i,\text{ Internal}}$, then $\sigma_{i,\text{ Max}} = \sigma_{i,\text{ Adjacent}}$, the artifacts near the central electrode are effectively eliminated. At the edge of the discrete phase, $\sigma_{i,\text{ Internal}} > \sigma_{i,\text{ Adjacent}}$, then $\sigma_{i,\text{ Max}} = \sigma_{i,\text{ Internal}}$, the clear contour of the discrete phase is maintained. However, the reconstructed values in the discrete phase near the central field may also be taken as the reconstructed values in the adjacent excitation mode. This makes the reconstructed conductivity in discrete phase higher, closer to 1, rather than 0, as shown in Fig. 10b.

The conductivity distribution under mixed excitation mode is analyzed as shown in Fig. 11. Compared with the internal electrode excitation method, the maximum reconstruction makes the reconstruction value of the discrete phase in the central region deviate from the true value, and its value is closer to 1.

In order to ensure that the reconstructed image has a clear contour and its internal conductivity is close to the true conductivity, the threshold can be used to judge the range of discrete phases in the new conductivity distribution σ_{Max}.

Setting threshold ρ, when the conductivity value of the i-th finite element is $\sigma_{i,\text{ Max}} > \rho$, we consider that the finite element is a continuous phase and adopt $\sigma_{i,\text{ Max}}$ as the conductivity of the finite element. When $\sigma_{i,\text{ Max}} < \rho$, the finite element is considered as discrete phase, and the average value $\sigma_{i,\text{ Mean}}$ of $\sigma_{i,\text{ Adjacent}}$ and $\sigma_{i,\text{ Internal}}$ (Eq. 7) can be calculated as the conductivity of the finite element.

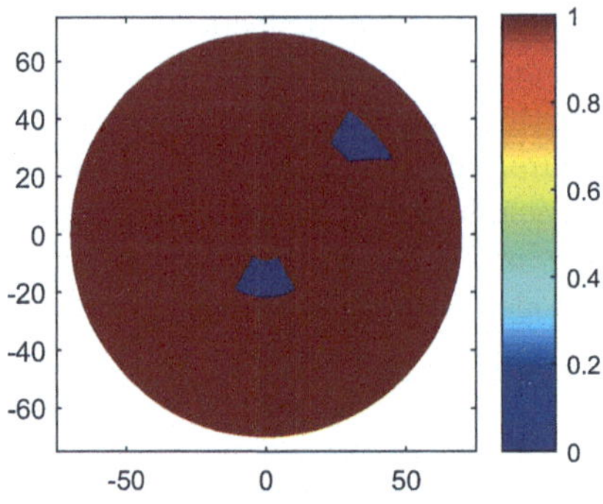

(a)

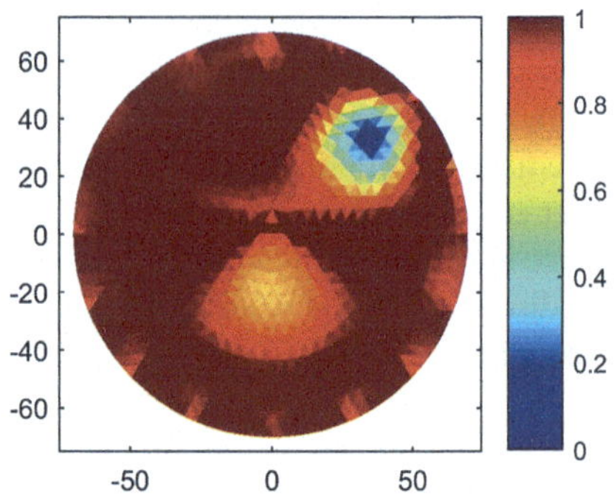

(b)

Fig. 10 Mixed excitation reconstruction using maximum value. (**a**) Model D1. (**b**) Reconstructed image of model D1

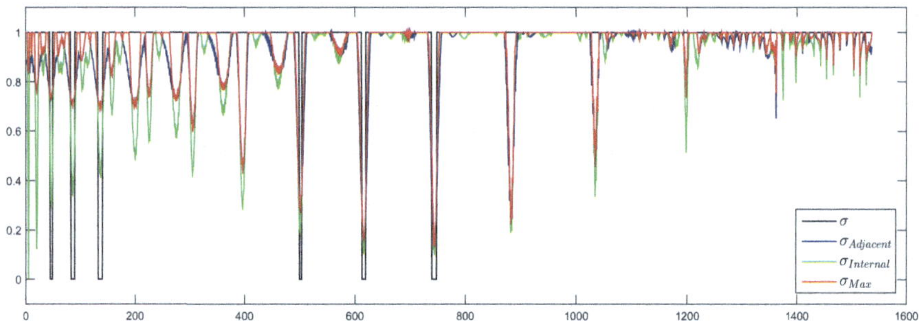

Fig. 11 Reconstructed conductivity amplitude

$$\sigma_{i,\text{Mean}} = \text{Mean}\left[\sigma_{i,\text{Adjacent}}, \sigma_{i,\text{Internal}}\right] \quad (i = 0, 1, \ldots, 1536) \tag{7}$$

The reconstruction algorithm can be expressed as

$$\sigma_{i,\text{Mix}} = \begin{matrix} \sigma_{i,\text{Max}} & \left(\sigma_{i,\text{Max}} > \rho\right) \\ \sigma_{i,\text{Mean}} & \left(\sigma_{i,\text{Max}} < \rho\right) \end{matrix}. \tag{8}$$

According to the experience, we choose $\rho = 0.7$ and use the algorithm to reconstruct the field as shown in Fig. 12b.

Comparing Figs. 12b and 10b, the image reconstructed by the new algorithm is closer to the real conductivity in the discrete phase near the central region.

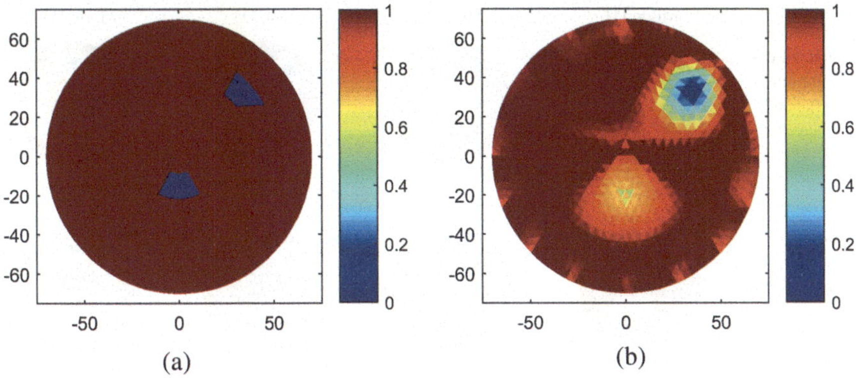

Fig. 12 Mixed excitation reconstruction using maximum value. (**a**) Model D1. (**b**) Reconstructed image of model D1

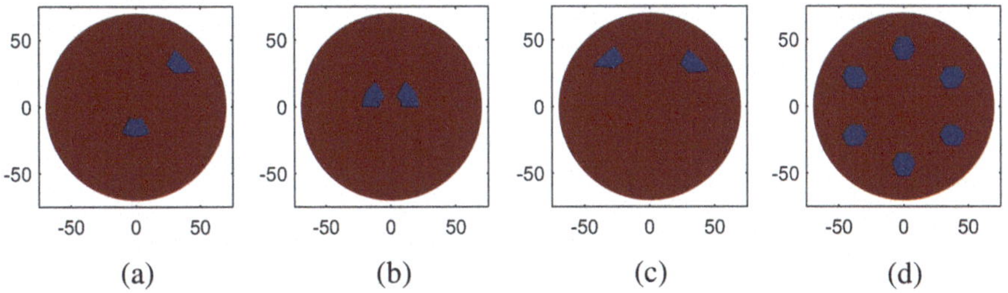

Fig. 13 Four conductivity distribution models for simulation. (**a**) Model D1. (**b**) Model D2. (**c**) Model D3. (**d**) Model D4

4 Simulation and Results

Experimental Conditions

In order to verify the reconstruction effect of internal electrode excitation mode for different distribution, four models are selected for simulation experiments as shown in Fig. 13.

Relative error of image reconstruction and image correlation coefficient are used to evaluate the reconstruction results [15].

$$\text{Reconstruction error} = \frac{\|\hat{\sigma} - \sigma\|}{\|\sigma\|} \times 100\% \tag{9}$$

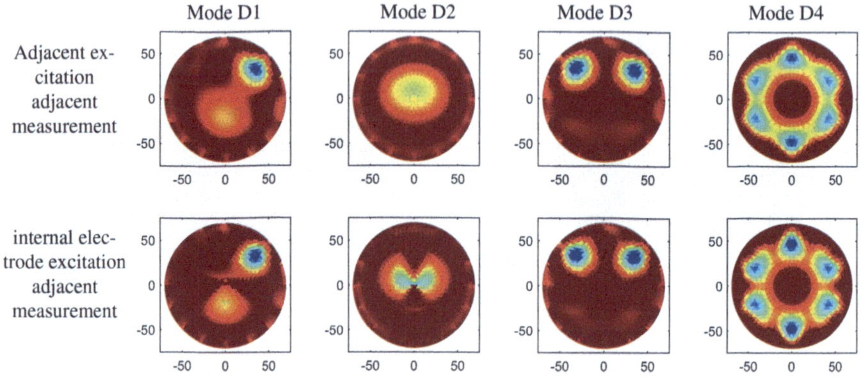

Fig. 14 Reconstruction of two excitation modes without noise

$$\text{Correlation coefficient} = \frac{\sum_{i=1}^{K} (\sigma_i - \overline{\sigma}) \left(\hat{\sigma}_i - \overline{\hat{\sigma}}\right)}{\sqrt{\sum_{i=1}^{K} (\sigma_i - \overline{\sigma})^2 \sum_{i=1}^{K} \left(\sigma_i - \overline{\hat{\sigma}}\right)^2}} \tag{10}$$

where $\hat{\sigma}$ is the calculated conductivity distribution in the field, σ is the true conductivity distribution, and $K = 1536$ is the number of finite elements in the field.

Noise Free

Four models are stimulated by the adjacent electrode excitation mode and the inner electrode excitation mode, respectively, and the Landweber algorithm is used to iterate 500 times for reconstruction. In the internal electrode excitation mode, the image is reconstructed by Eq. (8). The reconstructed image is shown in Fig. 14.

It can be seen that for different models, the reconstructed images using the proposed incentive mode are clear. The contour of discrete phase in the reconstructed image is clearer and the reconstruction of the central region of the field is more accurate. The image reconstruction errors and correlation coefficients are analyzed as shown in Fig. 15.

As shown in Fig. 15, for four different models, the reconstruction errors obtained by the proposed measurement model are reduced and the correlation coefficients are improved.

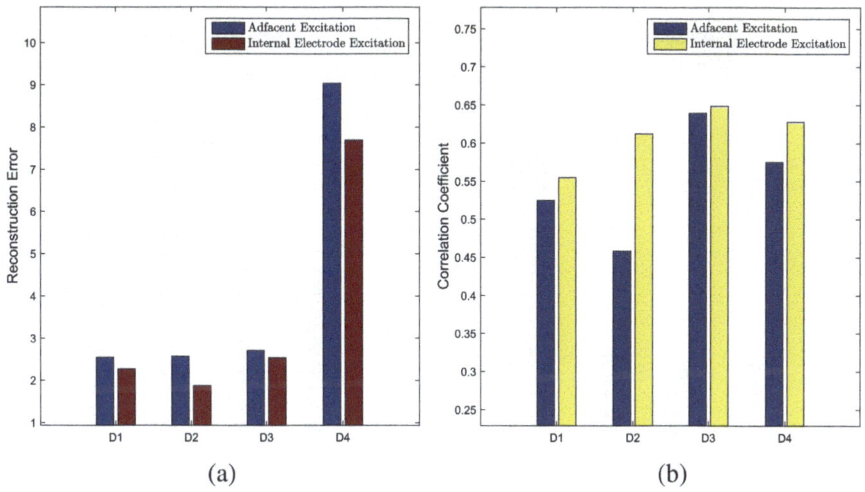

Fig. 15 Reconstruction errors and image correlation coefficients of different excitation. (**a**) Reconstruction errors. (**b**) Image correlation coefficients

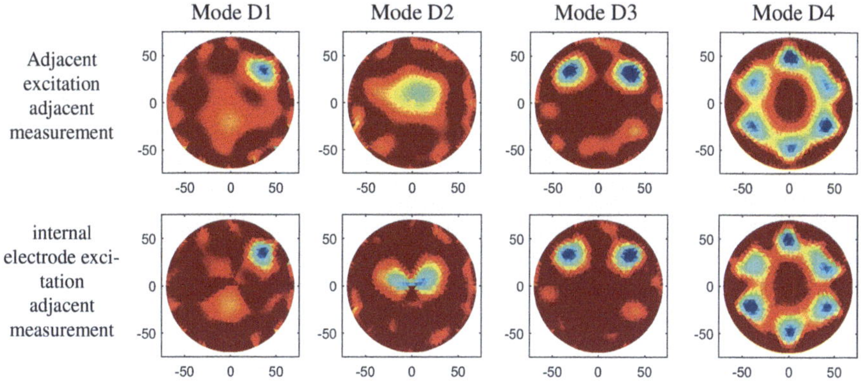

Fig. 16 Reconstruction of two excitation modes with noise

With Noise

In practical applications, the measured data are often disturbed by noise, so the anti-noise ability of the system is very important. In this chapter, by adding random noise to the boundary voltage vector, the SNR is 30 dB (Fig. 16).

Considering the system noise, the reconstructed image under adjacent excitation measurement mode is seriously distorted, especially near the central field. Using the excitation measurement mode proposed in this chapter, the discrete phases distributed in different positions can be reconstructed clearly by adding inner electrodes. The position of discrete phase distribution is basically correct and the

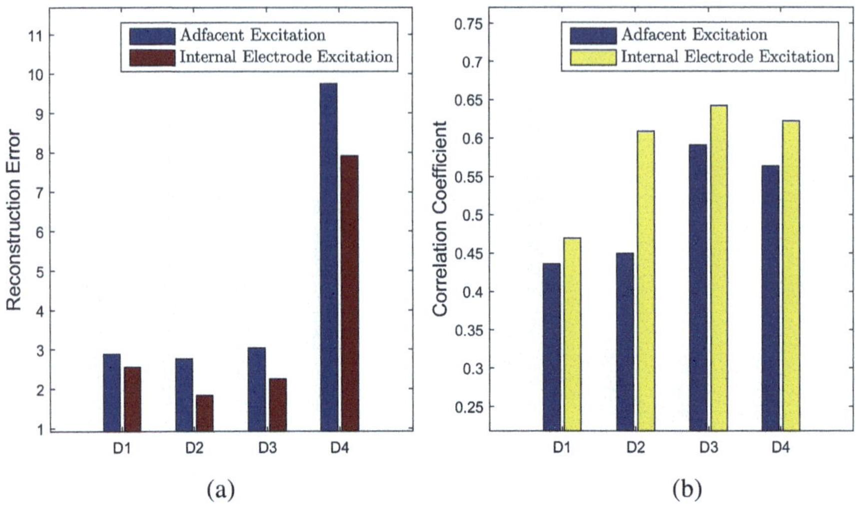

Fig. 17 Reconstruction errors and image correlation coefficients of different excitation with noise. (**a**) Reconstruction errors. (**b**) Image correlation coefficients

outline is clear. It can be considered that the anti-noise ability of the whole field has been improved obviously.

Quantitative analysis is made on the reconstructed image after adding noise. The reconstruction error and reconstruction correlation coefficient are shown in Fig. 17.

In the case of noise, for four different models, the reconstructed images obtained by the proposed excitation mode have lower image error than those measured by adjacent excitation. At the same time, the correlation coefficient between reconstructed distribution and real distribution has been effectively improved. Therefore, according to the quantitative analysis, we can conclude that the measurement mode calculation method proposed in this chapter has better anti-noise performance.

5 Conclusion

Based on the analysis of sensitivity coefficient, an ERT excitation measurement mode with inner electrodes is proposed, and the corresponding algorithm is studied. The simulation results show that the model can effectively reduce the reconstruction error and improve the correlation coefficient of the reconstructed image. However, it cannot be neglected that the internal electrode intrudes into the field, so the measurement mode proposed in this chapter cannot be applied to non-intrusive measurement environment. In addition, the proposed algorithm can be further improved and will be studied in future work.

References

1. Dickin, F., & Wang, M. (1996). Electrical resistance tomography for process applications. *Measurement Science and Technology, 7*, 247–260.
2. Kim, B. S., Khambampati, A. K., Kim, S., et al. (2015). Electrical resistance imaging of two-phase flow using direct Landweber method. *Flow Measurement and Instrumentation, 41*, 41–49.
3. Ye, J., Wu, M., Che, H., et al. (2016). Coupling Simulation for electrical resistance tomography. In *IEEE International Conference on Imaging Systems & Techniques*. IEEE.
4. Babaei, R., Bonakdarpour, B., & Ein-Mozaffari, F. (2015). The use of electrical resistance tomography for the characterization of gas holdup inside a bubble column bioreactor containing activated sludge. *Chemical Engineering Journal, 268*(3), 260–269.
5. Shao, Z., Wang, D., Wang, Y., et al. (2016). Electrical resistivity of coal-bearing rocks under high temperature and the detection of coal fires using electrical resistance tomography. *Geophysical Journal International, 204*(2), 1316–1331.
6. Andrade, R. (2011). Intervention of electrical resistance tomography (ERT) in resolving hydrological problems of a semi arid granite terrain of Southern India. *Journal of the Geological Society of India, 78*(4), 337–344.
7. Binley, A., Henrypoulter, S., & Shaw, B. (1996). Examination of solute transport in an undisturbed soil column using electrical resistance tomography. *Water Resources Research, 32*(4), 763–769.
8. Lehr, J. (1972). A vector derivation useful in impedance plethysmographic field calculations. *IEEE Transactions on Biomedical Engineering, 8*(4A), 156–157.
9. Liu, J. W., & Dong, F. D. F. (2004). Electrical resistance tomography based on the single drive electrode method. In *International Conference on Machine Learning & Cybernetics*. IEEE.
10. Ren, S., Dong, F., & Tan, C. (2011). High-precision electrical resistance tomography with external and internal electrode arrays. In *Instrumentation & Measurement Technology Conference*. IEEE.
11. Murai, T., & Kagawa, Y. (1985). Electrical impedance computed tomography based on a finite element model. *IEEE Transactions on Bio-Medical Engineering, 32*(3), 177–184.
12. Kotre, C. J. (1989). A sensitivity coefficient method for the reconstruction of electrical impedance tomograms. *Clinical Physics and Physiological Measurement, 10*(3), 275–281.
13. Yang, W. Q., & Peng, L. (2003). Image reconstruction algorithms for electrical capacitance tomography. *Measurement Science & Technology, 14*, R1–R13.
14. Wang, H.-x., & Cao, Z. (2006). Nonlinearity of "soft" field in electrical impedance tomography system—Based on statistical methods. *Journal of Tianjin University, 39*(5), 543–547.
15. Yang, W. Q., Spink, D. M., York, T. A., et al. (1999). An image-reconstruction algorithm based on Landweber's iteration method for electrical-capacitance tomography. *Measurement Science and Technology, 10*(11), 1065–1069.

Approximated Scale Space for Efficient and Accurate SIFT Key-Point Detection

Ying Wang, Yiguang Liu, Zhenyu Xu, Yunan Zheng, and Weijie Hong

Abstract The SIFT (scale invariant feature transform) key-point serves as an indispensable role in many computer vision applications. This paper presents an approximation of the SIFT scale space for key-point detection with high efficiency while preserving the accuracy. We build the scale space by repeated averaging filters to approximate the Gaussian filters used in SIFT algorithm. The accuracy of the proposed method is guaranteed by that an image undergoes repeated smoothing with an averaging filter is approximately equivalent to the smoothing with a specified Gaussian filter, which can be proved by the *center limit theorem*. The efficiency is improved by using integral image to fast compute the averaging filtering. In addition, we also present a method to filter out unstable key-points on the edges. Experimental results demonstrate the proposed method can generate high repeatable key-points quite close to the SIFT with only about one tenth of computational complexity of SIFT, and concurrently the proposed method does outperform many other methods.

Keywords SIFT · Key-point detector · Approximated Gaussian · Repeated averaging filters

1 Introduction

The SIFT [1] key-point plays an important role in computer vision and pattern recognition tasks such as structure from motion, object matching, object recognition, and texture analysis [2–6]. Though many other key-points have been proposed [7–11], the SIFT key-point is still favored by many applications especially for those with high accuracy needed [12]. However, the SIFT key-point has also been criticized for the drawback of its heavy computational burden, thus many variational

Y. Wang · Y. Liu (✉) · Z. Xu · Y. Zheng · W. Hong
College of Computer Science, Sichuan University, Chengdu, China
e-mail: liuyg@scu.edu.cn

E. T. Quinto et al. (eds.), *The Proceedings of the International Conference on Sensing and Imaging, 2018*, Lecture Notes in Electrical Engineering 606,
https://doi.org/10.1007/978-3-030-30825-4_3

methods have been proposed to accelerate the computational speed, but at the cost of insufficient accuracy or degraded repeatability of the key-points. To efficiently compute the SIFT key-point while preserving the accuracy and repeatability is the goal of this paper. The SIFT key-point is obtained by the DOG blob detector which is an approximation of LOG detector. There are also some other blob detectors such as [13–15], but their performance is inferior to SIFT in terms of accuracy and repeatability, as the *scale space theory*[16, 17] proved that the Gaussian kernel is the only smoothing kernel that could be used to create the image scale space. Based on the center limit theorem, repeated averaging filtering can be used to approximate Gaussian filtering. We propose a bank of averaging filters that accurately approximate the Gaussian filter in the SIFT algorithm, and the averaging filters are implemented via integral images. Because integral image is computed in our method, a combined method for eliminating key-points on the edge is also presented by reusing the integral images, which is more efficient than the original one in SIFT algorithm.

2 A Brief Review of SIFT Scale Space Construction and Key-Point Detection

The SIFT scale space consists of two pyramids, the Gaussian pyramid and the DOG(difference of Gaussian) pyramid. The Gaussian pyramid consists of M octaves and each octave consists of N layers. Suppose $L(a, b)$ denotes the ath layer in octave b, it is produced by the convolution of a Gaussian function with the input image $I(x, y)$

$$L(a, b) = G(x, y, \sigma) * I(x, y) \tag{1}$$

where $*$ stands for the convolution operation and

$$G(x, y, \sigma) = \frac{1}{2\pi\sigma^2} e^{-(x^2+y^2)/2\sigma^2} \tag{2}$$

where σ is determined by a, b and the scale of the first layer σ_0

$$\sigma = 2^{a+\frac{b}{N-3}} \cdot \sigma_0 \tag{3}$$

Since the computation complexity of Eq. (1) is proportion to σ^2, in practical implementation, smaller σ can be used due to the property of the Gaussian convolution that the convolution of two Gaussian functions is also a Gaussian with variance being the sum of the original variances. That is,

$$I(x, y) * G(x, y, \sigma_1) * G(x, y, \sigma_2) = I(x, y) * G(x, y, \sigma_3) \tag{4}$$

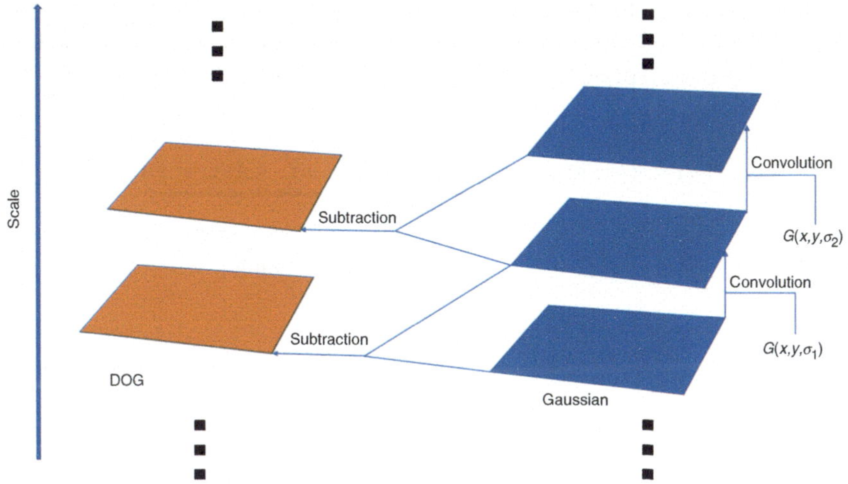

Fig. 1 The process to construct the SIFT scale space

where $\sigma_1^2 + \sigma_2^2 = \sigma_3^2$. So, if $b \neq 0$, $L(a, b)$ is computed as

$$L(a, b) = G(x, y, \sigma_{\text{gap}}) * L(a, b - 1) \tag{5}$$

where

$$\sigma_{\text{gap}} = \sqrt{\sigma(0, b)^2 - \sigma(0, b - 1)^2} \tag{6}$$

It can be seen from Eq. (6) that each octave shares the same group of σ to generate a successive layer, this is because when $b = 0$ and $a \neq 0$, the $L(a, b)$ is obtained by down sampling the layer $L(a - 1, b + N - 2)$. Once the Gaussian pyramid is computed, the difference of Gaussian pyramid is obtained via the subtraction of two successive layers in the Gaussian pyramid. Figure 1 shows the SIFT scale space construction processes. The SIFT key-point is detected by finding the extrema in the DOG pyramid, after that interpolation is performed to get sub-pixel precision if the key-point is not on the edge.

3 Approximation of Gaussian Smoothing Using Repeated Averaging Filter

In this section, we introduce the quantitative relationship between repeated averaging filtering and Gaussian filtering for a given image. The coefficients of an averaging filter can be viewed as the probability density of a random variable X

which is uniformly distributed in a square area centered at the origin. Given an image $I(x, y)$ and an averaging filter $A(x, y, w)$, where w is the window width of the averaging filter, repeated averaging filtering of the image can be mathematically expressed as:

$$I(x, y, n) = I(x, y) * \underbrace{A(x, y, w) * A(x, y, w) \ldots A(x, y, w)}_{n} \tag{7}$$

Based on the *center limit theorem*, when n approaches to infinity,

$$\begin{aligned}\lim_{n \to \infty} I(x, y, n) &= I(x, y) * \lim_{n \to \infty} \underbrace{A(x, y, w) * A(x, y, w) \ldots A(x, y, w)}_{n} \\ &= I(x, y) * G(x, y, \sigma_{\text{gau}})\end{aligned} \tag{8}$$

where $G(x, y, \sigma_{\text{gau}})$ is the Gaussian function with the variance σ_{gau}^2. Suppose the window width of an averaging filter is w, the variance of its corresponding discrete random variable is

$$\sigma_{\text{av}}^2 = \frac{w^2 - 1}{12} \tag{9}$$

the relation of σ_{av}, σ_{gau}, w and n can be deduced:

$$\sigma_{\text{gau}} = \sqrt{n\sigma_{\text{av}}^2} = \sqrt{\frac{nw^2 - n}{12}} \tag{10}$$

For discrete averaging filter and discrete Gaussian filter, the two sides of the first equal sign of Eq. (10) are very close to each other when $n \nless 3$, which indicates no less than 3 times repeated averaging filtering can well approximate a specified Gaussian filter. Figure 2 shows the two curves are close to each other.

4 Scale Space Construction via Repeated Averaging Filtering

To create the scale space introduced in Sect. 1, the key problem is to seek the initial scale σ_0, a group of Gaussian filters used to construct next layer from current layer and a group of averaging filters to approximate these Gaussian filters. Given a Gaussian kernel, it is not hard to find an optimal averaging filter $A(x, y, w)$ and the times n needed to approximate the Gaussian kernel based on Eq. (10). However, there are more difficulties to get the optimal averaging filter and filtering times in real applications. Here we mainly focus on finding optimal averaging filters to construct the scale space introduced in Sect. 1.

Fig. 2 Two similar curves: the red one is the Gaussian kernel with $\sigma = 2\sqrt{2}$, the blue one is its approximation resulted by 4 times convolution of an averaging filter with the window width $w = 5$

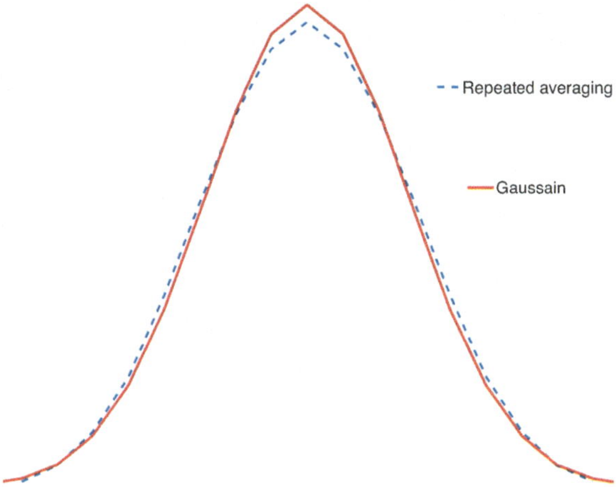

The constraint of Eq. (10) is that n is a positive integer not smaller than 3 and not bigger than 10, because if $n < 3$ it is not enough to approximate the Gaussian kernel and if $n > 10$ the computational complexity will have no advantage over the original Gaussian filtering. The variable w should be a positive odd number not smaller than 3, in order to satisfy that there could always be a center pixel the filtering result can be assigned to. Suppose the initial scale of the Gaussian pyramid is σ_0, as most key-points lie in the first several layer of the scale space, thus σ_0 should not be large. According to the original SIFT algorithm where $\sigma_0 = 1.6$, here we set $\sigma_0 < 2.0$. The number of layers N in each octave depends on the sampling frequency F, and $F = N - 3$. In [1] the sampling frequency is set to 2, 3, or 4, which gains an acceptable balance between key-point repeatability and computational time. Constraints to σ_i under SIFT scale space is that $\sigma_i = \sigma_0\sqrt{2^{\frac{2(i+1)}{F}} - 2^{\frac{2i}{F}}}$. Besides, $\sigma_i = \sqrt{\frac{n_i w_i^2 - n_i}{12}}$ is needed for discrete averaging filters and discrete Gaussian filters. These constraints can be formulated as:

$$\sigma_i = \sigma_0\sqrt{2^{\frac{2(i+1)}{F}} - 2^{\frac{2i}{F}}} = \sqrt{\frac{n_i w_i^2 - n_i}{12}} \tag{11}$$

where $i = 1, 2, 3, \ldots, F+2$, $\sigma_0 < 2$, $n_i \in \{3, 4, 5, \ldots, 10\}$, $w_i \in \{3, 5, 7, \ldots\}$. It can be verified that the only solution to Eq. (11) is

$$\begin{cases} \sigma_0 = \sqrt{2}, \sigma_1 = 2, \sigma_2 = 2\sqrt{2}, \sigma_3 = 4, \sigma_4 = 4\sqrt{2} \\ n_1 = 3, n_2 = 6, n_3 = 4, n_4 = 4 \\ w_1 = 3, w_2 = 3, w_3 = 5, w_4 = 7 \\ F = 2 \end{cases} \tag{12}$$

Fig. 3 Approximated DOG detector for SIFT key-point detection

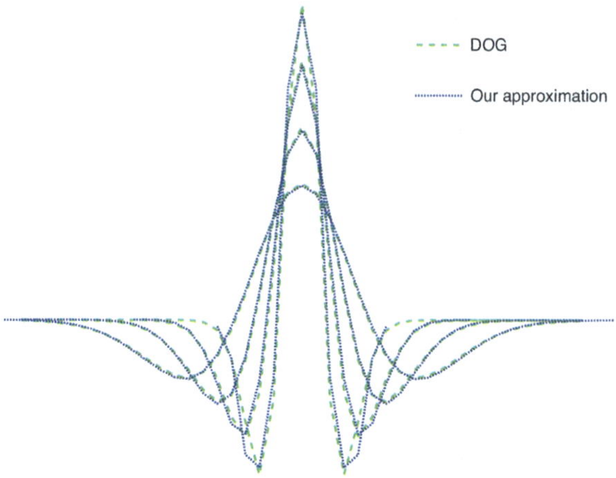

The approximated SIFT scale space is constructed in the same procedure as it is in the original SIFT algorithm, the Gaussian filters are replaced by averaging filters, and their quantitative relation is given in Eq. (12). Note that the key-point detector in SIFT is actually the DOG detector obtained from subtraction of two Gaussian kernels, Fig. 3 shows the approximated DOG and DOG curves in one octave of the DOG pyramid.

5 Key-Point Detection

The key-points are detected by searching the local extrema in a 3×3 region of the approximated DOG pyramid. However, the DOG detector is sensitive to image edges, and the key-point on the edge should be removed since it is unstable. As integral image is computed in our method (depicted in Fig. 4a), we propose to use fast Hessian measure to reject key-points on the edge. The principal curvature is used to determine whether a point is on the edge, which is computed by the eigenvalues of scale adapted Hessian matrix:

$$H(X, \sigma) = \begin{bmatrix} L_{xx}(X, \sigma), & L_{xy}(X, \sigma) \\ L_{xy}(X, \sigma), & L_{yy}(X, \sigma) \end{bmatrix}, \tag{13}$$

The partial derivatives in Eq. (13) are computed by integral image and a group of box filters as illustrated in Fig. 4. The images of before and after filtering out the key-point on the edge are shown in Fig. 5. After filtering out the key-points on edge, the scale space quadratic interpolation is performed for every key-point to get a sub-pixel accurate location (Fig. 5).

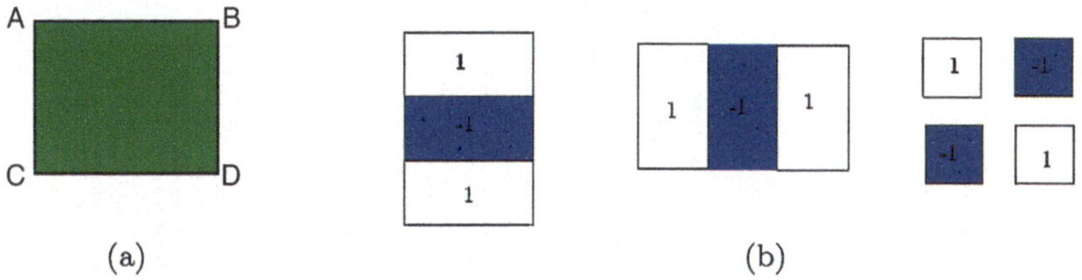

Fig. 4 Using integral image and box filters to compute the partial derivatives of Hessian matrix. (**a**) The integral image: $SUM = A + D - B - C$. (**b**) Box filters to compute the partial derivatives of the Hessian matrix

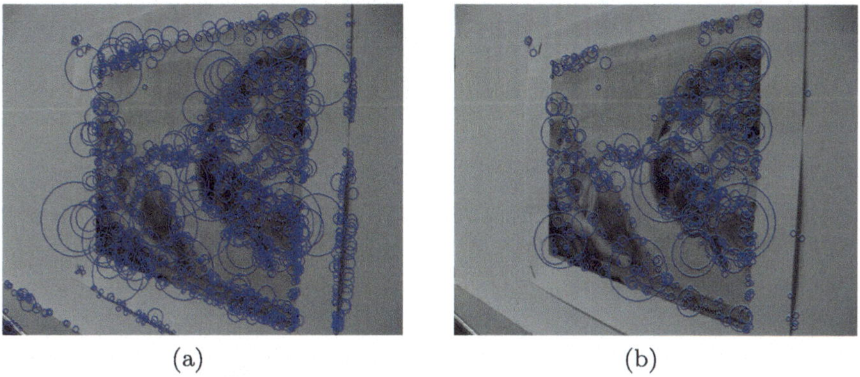

Fig. 5 Before-and-after filtering out the edge key-points. (**a**) Before filtering out the edge key-points. (**b**) After filtering out the edge key-points

6 Experimental Results

The accuracy of our method is validated by the performance comparison of the approximated SIFT detector, SIFT detector, SURF detector, and FAST detector. The SIFT code implemented by Rob Hess [18] is used in our experiment. For the SURF algorithm, we use the original implementation released by its authors. The FAST detector in our experiment is implemented in OpenCV 3.2.0. The datasets proposed by Mikolajczyk and Schmid [19] are used for the evaluation. There are totally 8 datasets, and each dataset has six images with increasing amount of deformation from a reference image. The deformations, covering zoom and rotation (Boat and Bark sequences), view-point change (Wall and Graffiti sequences), brightness changes (Leuven sequence), blur (Trees and Bikes sequence) as well as JPEG compression (UBC sequence), are provided with known ground truth which can be used to identify the correspondence. The *repeatability score* introduced in [19] is used to measure the reliability of the detectors on detecting the same feature point under different deformations of the same scene. The repeatability score is defined as the ratio between the number of corresponding features and the smaller number of features in one image in the image pair. To give a fair comparison, the thresholds of these detectors are adjusted to give approximately equal number of feature points

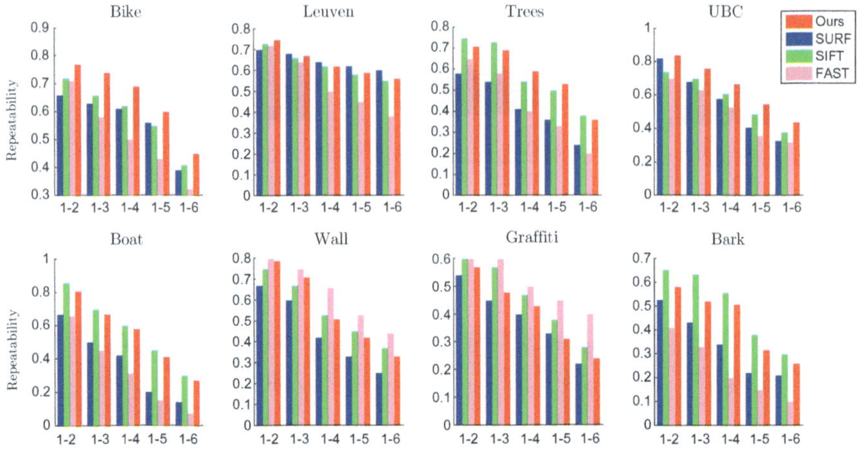

Fig. 6 The repeatability of several key-point detectors

Table 1 Timing results of SIFT, SURF, FAST and our method for key-point detection on the first image of the Wall sequence (size: 1000 × 700 pixels)

Detector	SIFT	SURF	FAST	OURS
Time (ms)	332	97	11	35

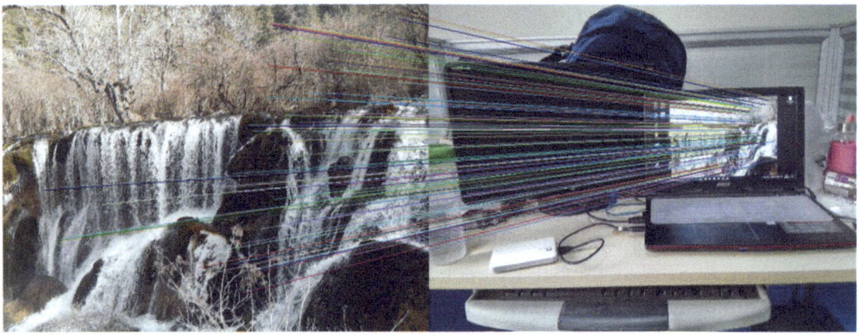

Fig. 7 Real-world image matching using the approximated SIFT detector with SURF descriptor

as SURF detector detected. The results in Fig. 6 show that the SIFT detector has the best performance in most cases, and the proposed method is close to SIFT and better than SURF and FAST detector.

The efficiency of the proposed method is also compared to SIFT, SURF, and FAST. The times of detection of the key-points in the first image of the wall sequence is shown in Table 1, where we can see that the FAST detector is most efficient among all methods, but it merely detects key-points in one scale, while others detect them in more than 10 scales. The proposed method is about 3 times faster than SURF and 10 times faster than SIFT. Real-world image matching using the approximated SIFT detector with SURF descriptor is shown in Fig. 7.

7 Conclusion

In this paper we have proposed a framework to efficiently and accurately approximate the SIFT scale space for key-point detection. The quantitative relation between repeated averaging filtering and Gaussian filtering has been analyzed, and a group of averaging filters has been found to accurately approximate the DOG detector. Experimental results demonstrate that the proposed method is about 10 times faster than SIFT detector, while preserving the high accuracy and high repeatability of the SIFT detector.

Acknowledgements This work is supported by NSFC under Grant 61860206007 and 61571313, by National Key Research and Development Program under Grant 2016YFB0800600 and 2016YFB0801100 and by funding from Sichuan Province under Grant 18GJHZ0138, and by joint funding from Sichuan University and Lu-Zhou city under 2016CDLZ-G02-SCU.

References

1. Lowe, D. G. (2004). Distinctive image features from scale-invariant keypoints. *International Journal of Computer Vision, 60*(2), 91–110.
2. St-Charles, P. L., Bilodeau, G. A., & Bergevin, R. (2016). Fast image gradients using binary feature convolutions. In *Proceedings of the IEEE Conference on Computer Vision and Pattern Recognition Workshops* (pp. 1–9).
3. Loncomilla, P., Ruiz-del-Solar, J., & Martnez, L. (2016). Object recognition using local invariant features for robotic applications: A survey. *Pattern Recognition, 60*, 499–514.
4. Ma, W., Wen, Z., Wu, Y., Jiao, L., Gong, M., Zheng, Y., et al. (2016). Remote sensing image registration with modified SIFT and enhanced feature matching. *IEEE Geoscience and Remote Sensing Letters, 14*(1), 3–7.
5. Yang, C., Wanyu, L., Yanli, Z., & Hong, L. (2016). The research of video tracking based on improved SIFT algorithm. In *2016 IEEE International Conference on Mechatronics and Automation (ICMA)* (pp. 1703–1707). Piscataway, NJ: IEEE.
6. Stcharles, P. L., Bilodeau, G. A., & Bergevin, R. (2016). Fast image gradients using binary feature convolutions. In *IEEE Computer Society Conference on Computer Vision and Pattern Recognition Workshops* (pp. 1074–1082). Piscataway, NJ: IEEE.
7. Bay, H., Ess, A., Tuytelaars, T., & Van Gool, L. (2008). Speeded-up robust features (SURF). *Computer Vision and Image Understanding, 110*(3), 346–359.
8. Calonder, M., Lepetit, V., Strecha, C., & Fua, P. (2010). Brief: Binary robust independent elementary features. In *European Conference on Computer Vision* (pp. 778–792). Berlin: Springer.
9. Rublee, E., Rabaud, V., Konolige, K., & Bradski, G. R. (2011). ORB: An efficient alternative to SIFT or SURF. In *2011 IEEE International Conference on Computer Vision (ICCV)* (pp. 2564–2571). Piscataway, NJ: IEEE.
10. Leutenegger S, Chli M, Siegwart R. Y. (2011). BRISK: Binary robust invariant scalable keypoints. In *2011 IEEE International Conference on Computer Vision (ICCV)* (pp. 2548–2555). Piscataway, NJ: IEEE.
11. Alahi, A., Ortiz, R., & Vandergheynst, P. (2012). Freak: Fast retina keypoint. In *2012 IEEE Conference on Computer Vision and Pattern Recognition (CVPR)* (pp. 510–517). Piscataway, NJ: IEEE.

12. Mitsugami, I. (2013). Bundler: Structure from motion for unordered image collections. *Journal of the Institute of Image Information & Television Engineers, 65*(4), 479–482.
13. Agrawal, M., Konolige, K., & Blas, M. R. (2008). CenSurE: Center surround extremas for realtime feature detection and matching. In European Conference on Computer Vision (ECCV) (pp. 102–115). Berlin: Springer.
14. Matas, J., Chum, O., Urban, M., & Pajdla, T. (2004). Robust wide-baseline stereo from maximally stable extremal regions. *Image and Vision Computing, 22*(10), 761–767.
15. Grabner, M., Grabner, H., & Bischof, H. (2006). Fast approximated SIFT. In *Asian Conference on Computer Vision* (pp. 918–927). Berlin: Springer.
16. Crowley, J. L., & Parker, A. C. (1984). A representation for shape based on peaks and ridges in the difference of low-pass transform. *IEEE Transactions on Pattern Analysis and Machine Intelligence, 6*(2), 156–170.
17. Lindeberg, T. (1994). Scale-space theory: A basic tool for analysing structures at different scales. *Journal of Applied Statistics, 21*(2), 225–270, 21(2):224–270.
18. Hess, R. (2010). An open-source SIFTLibrary. In *Proceedings of the Inter National Conference on Multimedia* (pp. 1493–1496). New York, NY: ACM.
19. Mikolajczyk, K., Tuytelaars, T., Schmid, C., Zisserman, A., Matas, J., Schaffalitzky, F., et al. (2005). A comparison of affine region detectors. *International Journal of Computer Vision, 65*(1–2), 43–72.

The Application of Image Processing in UAV Reconnaissance Information Mining System

Qiufeng Pan, Jiaxing Wang, Hang Yu, Wenjun Zhang, and Peng Yue

Abstract As a kind of emerging weapon, UAV (unmanned aerial vehicle) has become increasingly prominent in modern warfare and received unprecedented attention from the world's military powers. The major function of UAV in warfare is reconnaissance, providing intelligence support for systematic warfare with the equipped different kinds of payloads. Videos and images are the most intuitive reconnaissance results and the basic carrier of the reconnaissance information processing system. This paper analyzes the application of image processing technology in UAV intelligence information processing through demand analysis of reconnaissance mission-oriented and the procedure of reconnaissance mission. We finally complete the design and implementation of UAV reconnaissance information processing system.

Keywords Emerging weapon · Unmanned aerial vehicle · Image processing · Intelligence support · Reconnaissance information mining system

1 Introduction

Modernized war deeply depends on the battlefield intelligence information. Ground reconnaissance equipment would be influenced by the obstacle and earth curvature. Satellite reconnaissance equipment couldn't detect in real time and the details are not clear. Thus, these two reconnaissance methods have their own limits. Reconnaissance UAV is an important complementary to the other two detective means [1]. With different payloads carried, reconnaissance UAV could detect the details of battlefield situation in real time [2]. In the prewar period, reconnaissance of UAV could construct the battlefield situation, such as the map of battlefield, helping the commander make correct decision about the forces deployment. In

Q. Pan · J. Wang (✉) · H. Yu · W. Zhang · P. Yue
Research Laboratory of Command and Control and System Simulation Technology, Beijing Aerospace UAV System Engineering Research Institute, Beijing, China

E. T. Quinto et al. (eds.), *The Proceedings of the International Conference on Sensing and Imaging, 2018*, Lecture Notes in Electrical Engineering 606,
https://doi.org/10.1007/978-3-030-30825-4_4

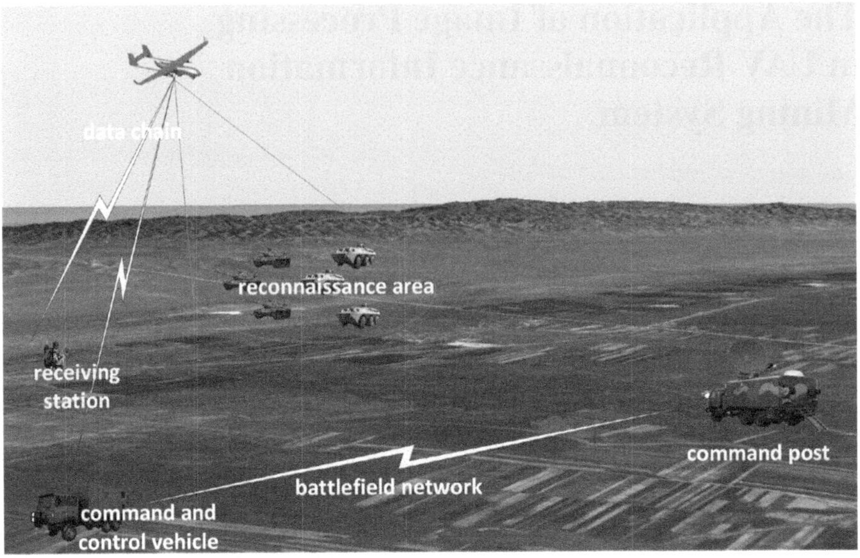

Fig. 1 UAV participated in the war

the wartime, reconnaissance of UAV could detect the change of battlefield, detect and recognize target, helping the commander revise war policy in time [3]. In the postwar period, the damage assessment is an important evidence of summary [4]. The major form of reconnaissance is to capture images and videos from the battle field. The image acquired by the payload is an intuitive intelligence source. Thus, image processing technology directly influenced the intelligence result.

Figure 1 shows the process of UAV participation in the war. The UAV is reconnoitering the battlefield situation, sending the reconnaissance results to the receiving station and command and control vehicles by data chain in real time. The intelligence processing system is installed on the vehicle to generate the intelligence information and report the information to command post to help the commander make decisions about how to deploy forces by battlefield network.

This paper is organized as follows. Section one educes the importance of image processing in intelligence mining system with the way UAV is providing the intelligence support in war. Section two analyzes the requirement analysis of image processing technology. Section three describes the implementation of the used image processing technology. Section four concludes this paper.

2 Requirements Analysis

The UAV reconnaissance information processing system integrates the intelligence information obtained by various means during the reconnaissance task. The function of this system contains intelligence generation, sharing and distribution, and intelligence support throughout the process of mission execution. It is the information analysis and processing center of UAV reconnaissance mission system.

The requirements of UAV reconnaissance intelligence include two aspects. One is obtaining the high-precision reconnaissance panorama image. The other is detection, recognition, and location of enemy target. To obtain high-precision and large-scale images in the reconnaissance area to help construct the battlefield situation in real time, image mosaic method is applied. After image mosaic, combined with the map of the battlefield such as "Google Global Satellite Map," we can get the location of target in the panoramic image. The location and property of targets are reported to the command post to decide whether to attack. Then artillery correction technology is used to amend the fire attack system if the fire did not hit the target. Also, if the target is hit by artillery, the damage assessment could help commander to summary the war to determine whether the target is damaged. Meanwhile, in most cases, the battlefield environment is complex. Single mode cameras may not be able to detect targets. SAR (synthetic aperture radar) or infrared camera may be also used to detect target.

3 Key Technology

In this section, we discuss the image processing technology in the intelligence mining system, containing (1) image mosaic and geometric correction, (2) image fusion, (3) target detection and recognition, (4) target location, and (5) artillery correction and damage assessment.

Image Mosaic and Electronic Map Fusion

After the mission is distributed to UAV, it flies to the mission area to execute reconnaissance mission. Because of the miniaturization of UAV and the limitations of the reconnaissance equipment, the scene structure information in a single frame image is very limited, which is not sufficient to obtain wide-area reconnaissance information. Thus we need to stitch the continuous multi-frame images to form a panoramic image. The panoramic image could also fusion with the electronic map to form the battlefield situation through coordinate transformation so that we could locate any point in the image in the world coordinate system [5]. The whole process of the algorithm is shown in Fig. 2.

The algorithm divides into two parts, image mosaic and geometric rectification.

1. Image Mosaic: Firstly, we screened the image data. Due to the sensor equipment trouble or the influence of data chain interference, the image data acquired by

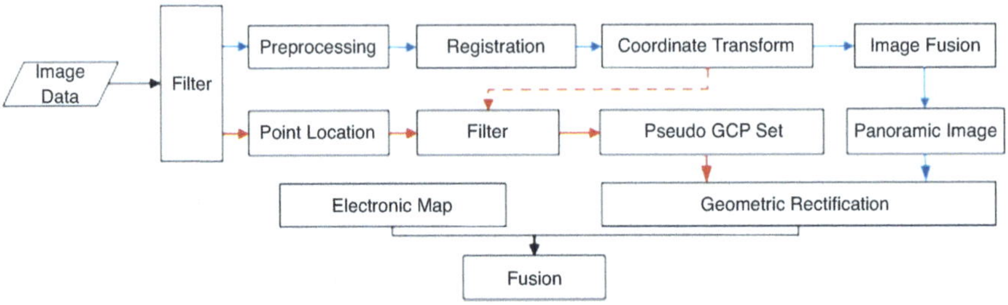

Fig. 2 Flowchart of building battlefield situation

UAV may contain some outlier data. The payload on UAV captures images about 25 frames per second, resulted in the large overlap region in the adjacent images, which means massive repeated information is contained in the image data. Considering the effectiveness of image data and the overlap region between images, several images are selected to stitch the panoramic image, such as six or seven images per second. Image mosaic is to transform all the images into a uniform coordinate. The first step is to register the image. Keypoint based method is fast and robust way to register images, containing keypoint extraction such as SIFT, SURF, ORB, and so on [6–8], construction of descriptor, and keypoint match. According to the matched keypoint, the transformation between images is calculated. Let the first image to be the benchmark and all the images could be transformed to the same coordinate.

2. When the panoramic image is acquired, according to the flight parameters of UAV, we could calculate the coordinate of the corner points on images in WGS84 (world geodetic system 1984), which is explained in the subsection target location. These corner points are used to be the pseudo GCP (ground control point) set to simulate the geometric distortion of images. The transformation model is based on the quadratic polynomial model, and the mathematical transformation relation is as follows.

$$\begin{cases} x = a_0 + (a_1 X + a_2 Y) + \left(a_3 X^2 + a_4 XY + a_5 Y^2\right) \\ y = b_0 + (b_1 X + b_2 Y) + \left(b_3 X^2 + b_4 XY + b_5 Y^2\right) \end{cases} \tag{1}$$

(x, y) and (X, Y) are the coordinates of arbitrary pixels in original and corrected images. (a_i, b_i), $(i=0, 1, \ldots, 5)$ are the coefficients of this model, which could be obtained by the least square method.

Figure 3 shows the fusion result of panoramic image and electronic map [9].

Visual and Infrared Image Fusion

The reconnaissance UAV carried a photoelectric payload including a visual camera and an infrared camera. The fusion between visual and infrared image could extract

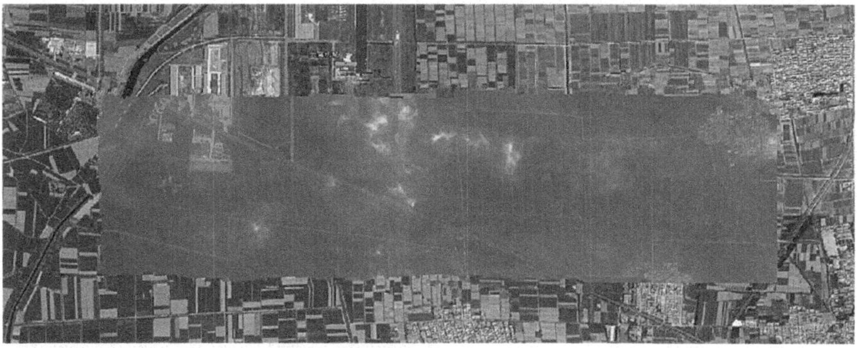

Fig. 3 Fusion result of panoramic image and electronic map

the most useful information of the two images, integrating them into a high-quality image. This effort could improve the utilization rate of image information.

Firstly, as the field of visual camera and infrared camera are not the same. We need to extract the overlap region of visual and infrared image to fuse. While the visual camera and infrared camera are in one pod, these two cameras would be focused on the same place, so that the center of visual image and infrared image are the same. According to the camera parameters, focal length and pixel size, we could easily extract the overlap region between visual camera and infrared camera.

Now, there are two of the most commonly used ways to integrate visual and infrared image: pixel level image fusion and feature level image fusion. Pixel level image fusion are fused on the original data level, contains more abundant details, while feature level image fusion extract features in the two images such as shapes, edge, contour, regions, etc. Thus feature level image fusion not only keeps effective information, but also compresses information.

In our system, the pixel level image fusion is utilizing the weighted mean of the luminance component of visual image and infrared image.

As is known to us, the background of infrared image is smooth and the difference between feature and background is in brightness. Thus, the most important thing is to extract the features of infrared image. We firstly utilize quad-tree decomposition to divide the infrared image and select the points in each image patch uniformly to be the control point of Bessel interpolation to reconstruct the background of infrared image. This method would result in the blocking artifact because of the different control points used in the stitched surface. Thus, the reconstructed background should be smoothed by a Gaussian filter to eliminate the blocking artifact. With the background of infrared image, we could easily obtain the features of the infrared image.

Figure 4 shows the fusion result of visual and infrared image with pixel level fusion and feature level fusion.

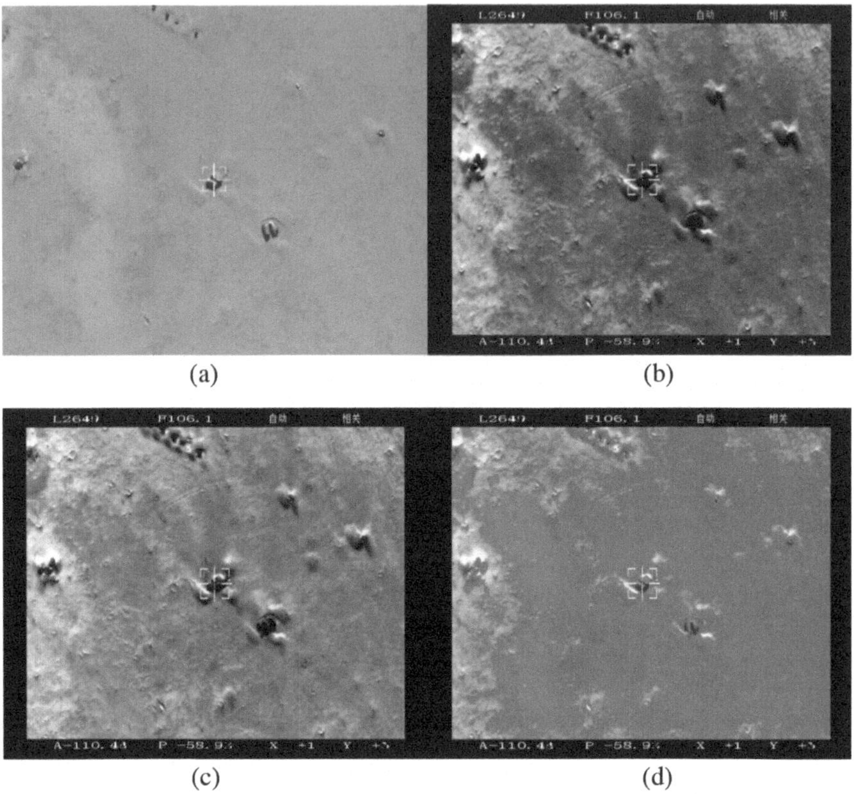

Fig. 4 Pixel level image fusion result and feature level image fusion result. (**a**) is visual image; (**b**) is infrared image; (**c**) is pixel level image fusion result; and (**d**) is feature level image fusion result

Target Detection and Recognition

Object detection system can detect and recognize targets on visible images, SAR images, and infrared images. The whole process is shown in Fig. 5.

In the object detection on visual images, deep learning has become one of the most popular methods in recent years [8]. It is a brain-like architecture that can be trained with massive data to extract features and learn the key features in sight of these data automatically. Deep learning network has deeper network than normal machine learning model and so that it can handle with more complicated features and realize object detection with high accuracy (Fig. 6).

In the object detection on SAR and infrared images, which have only one channel, denoising, thresholding, and dilating are used to make objects much prominent from the background. Then use the convolution filter to sample the objects and filter the boxes with non-maximum suppression to get the accurate location of objects (Fig. 7).

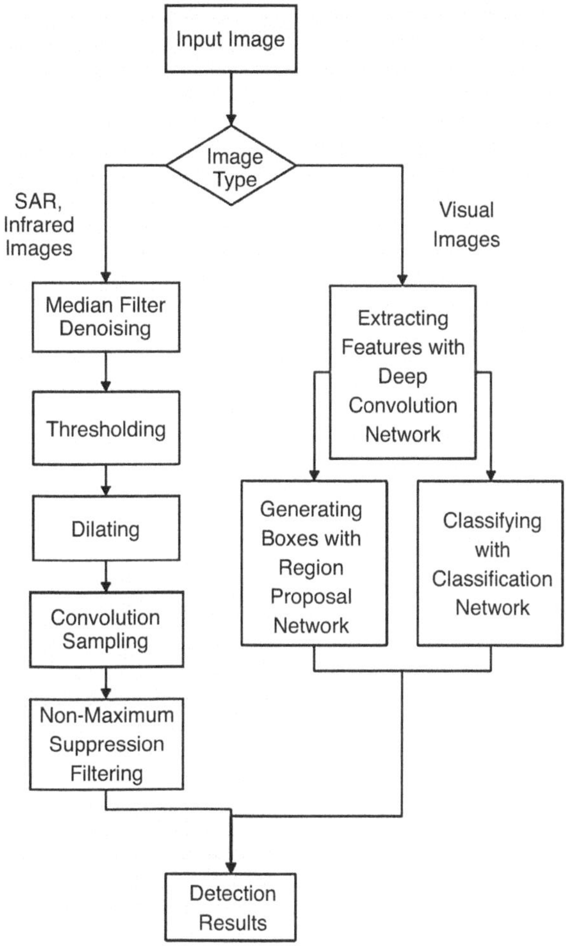

Fig. 5 The flowchart of target detection and recognition

Target Location

After the target is detected and recognized, the location of target is an important property used to guide the fire attack system to attack the target. The coordinate system we utilized is WGS 84 (World Geodetic System 1984). The relevant method is target location. It is based on photogrammetry and image processing technology, and it aims to calculate the latitude and longitude of target with the flight parameters of UAV in our system.

There are mainly three methods to locate the target.

The first one is based on the telemetry data of UAV. It could locate the target in real time while the location accuracy is not enough. The real time location of UAV is measured in time by the GPS (global position system) mounted in the UAV. With the altitude of UAV and elevation data in DEM format, the distance between UAV

Fig. 6 Detection results on visual images

Fig. 7 Detection results on SAR images

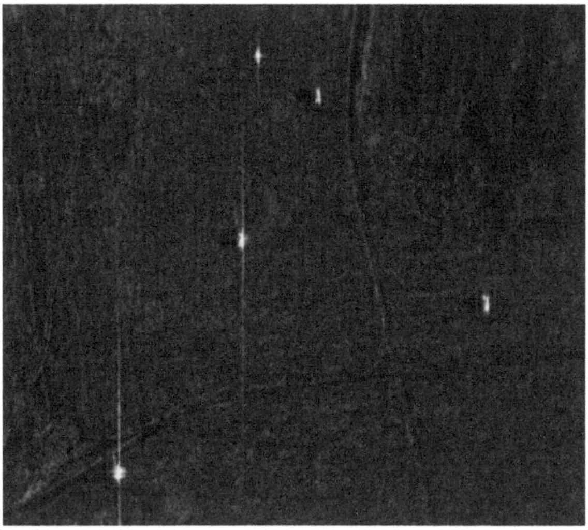

and ground is calculated. Also, the position, posture, azimuth, and pitch angle are measured by the sensor mounted on the UAV in real time.

According to the principles of imaging, the point location formula is as follows:

$$\begin{cases} X_A - X_s = (Z_A - Z_S)(a_1 \times x + a_2 \times y - a_3 \times f)/(c_1 \times x + c_2 \times y - c_3 \times f) \\ Y_A - Y_s = (Z_A - Z_S)(b_1 \times x + b_2 \times y - b_3 \times f)/(c_1 \times x + c_2 \times y - c_3 \times f) \end{cases} \quad (2)$$

(X_A, Y_A, Z_A) is the coordinate of object point A in ground photogrammetric coordinate system, and (X_S, Y_S, Z_S) is the coordinate of photography center. $(x, y, -f)$ is the coordinate of image point a in image space coordinate system S_XYZ. f is the focal length. a_1, a_2, a_3, b_1, b_2, b_3, c_1, c_2, c_3 are the elements of rotation matrix R. The rotation matrix R could be calculated as follows

$$R = \begin{bmatrix} a_1 & a_2 & a_3 \\ b_1 & b_2 & b_3 \\ c_1 & c_2 & c_3 \end{bmatrix}$$
$$= \begin{bmatrix} \cos k & \sin k & 0 \\ -\sin k & \cos k & 0 \\ 0 & 0 & 1 \end{bmatrix} \begin{bmatrix} \cos\omega & 0 & -\sin\omega \\ 0 & 1 & 0 \\ \sin\omega & 0 & \cos\omega \end{bmatrix} \begin{bmatrix} 1 & 0 & 0 \\ 0 & \cos\varphi & -\sin\varphi \\ 0 & \sin\varphi & \cos\varphi \end{bmatrix} \quad (3)$$

φ, ω, k are the pitch, roll, and yaw of payload in $\varphi - \omega - k$ rotation system.

The second method is based on the space intersection. It is the extension of the first method. Laser range finder is used in this method. It is used to measure the distance between target and UAV relatively accurately. With the location result when the UAV is locating the same target at different place, we can construct the intersection model. Least-square method could be used to reduce the location errors. Since the laser range finder is a more stable and accurate device, the measured result could be used as a benchmark value.

The formula is as follows:

$$(x, y, z) = \min \sum \left(f\left(x_n, y_n, z_n\right) - L_n\right)^2 \quad (4)$$

(x, y, z) is the location result. $f(x_n, y_n, z_n)$ is the distance between UAV and target calculate by the first method in time n. L_n is the measure esult of laser range finder in time n. Compared with the first method, the method with space intersection model is a more accurate way to locate the target.

The third method is based on the map match. It is also the extension of the first method and is similar to geometric rectification. From the first method, we can calculate the latitude and longitude of the visual image's four vertex to superpose the visual image over "Google Global Satellite Map" with some errors. Then, register the visual image and map image automatically or manually. Figure 8 shows the result of map matching between visual image and the relative image in "Google Global Satellite Map." The first image is the visual image captured by the payload. The figure shows the visual image superimposed over the "Google Global Satellite Map."

Artillery Correction and Damage Assessment

After the target property and location are sent to the command post, fire attacking system decides to attack the detected target. UAV would hover over the target to monitor it to know the strike situation by comparing the images before and after

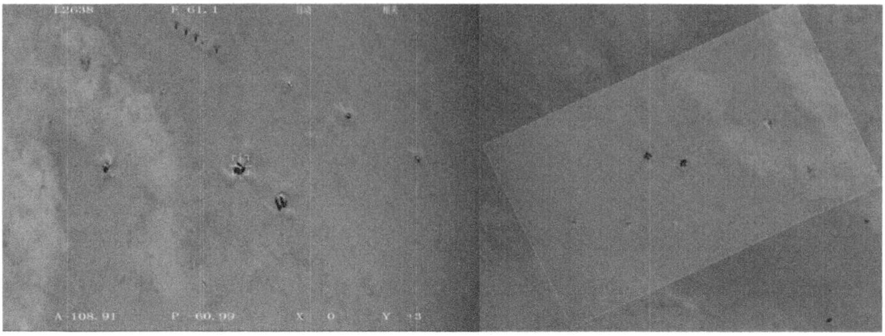

Fig. 8 location based on map match

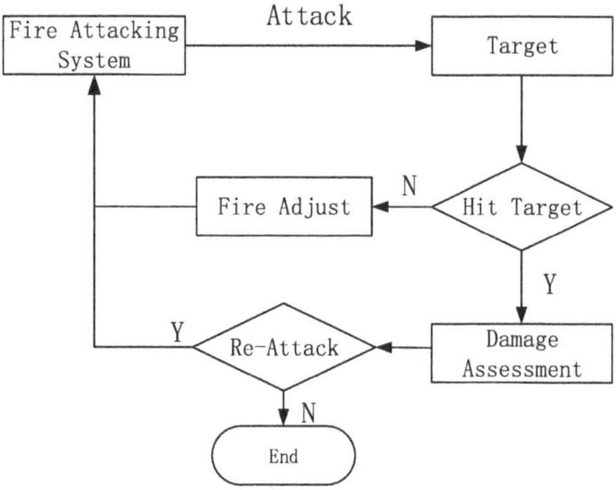

Fig. 9 The flowchart of artillery correction and damage assessment

attack. The process is shown in Fig. 9. Artillery correction is to calculate the distance and angle between the location of target and the impact point. By sending the distance and angle of impact point relative to the target to the fire attacking system, they can amend system to attack again until the target is hit. After the target is hit, the damage image is captured. By comparing the images before and after attacking, according to the attributes of the target, type and killing radius of the warhead, and other features, the damage situation is assessed to decide whether to attack again. The result can be seen in Fig. 10.

4 Conclusion

UAV intelligence information processing method is a branch of UAV combat application research. Many mature technologies in image processing will be applied in the information processing of UAV, such as moving target tracking, removing

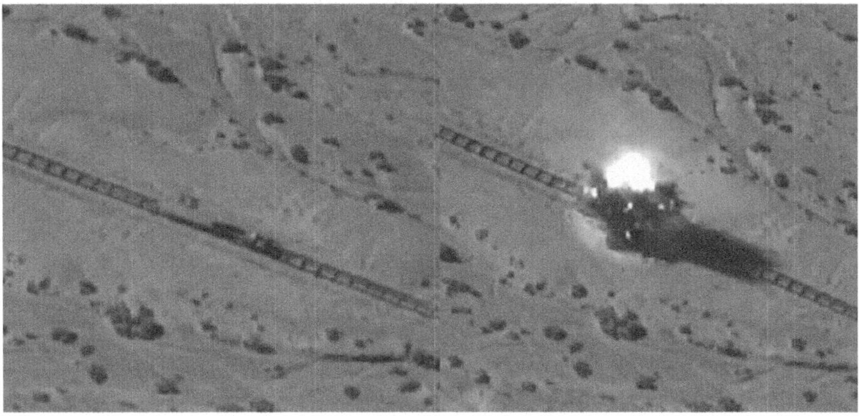

(a) (b)

Fig. 10 Comparison before and after attack. (**a**) Before attack. (**b**) After attack

fog or rain effect from images, and so on. With the improvement of the technical level of the UAV and the diversification of the reconnaissance payload, the ability of intelligence mining system also needs to improve. Therefore, image processing method will also face many new challenges.

References

1. Romaya, B. (2018). Transformations of war: Philosophical challenges in contemporary times. In *Proceedings of the XXIII World Congress of Philosophy* (Vol. 69).
2. Ashkenazi, M. (2017). The future of UAVs: Lessons from the 'Great War'. *S&F Sicherheit und Frieden, 34*(4), 257–262.
3. Laliberte, A. S., et al. (2010). Acquisition, orthorectification, and object-based classification of unmanned aerial vehicle (UAV) imagery for rangeland monitoring. *Photogrammetric Engineering & Remote Sensing, 76*(6), 661–672.
4. Fernandez Galarreta, J., Kerle, N., & Gerke, M. (2015). UAV-based urban structural damage assessment using object-based image analysis and semantic reasoning. *Natural Hazards and Earth System Sciences, 15*(6), 1087–1101.
5. Turner, D., Lucieer, A., & Watson, C. (2012). An automated technique for generating georectified mosaics from ultra-high resolution unmanned aerial vehicle (UAV) imagery, based on structure from motion (SfM) point clouds. *Remote Sensing, 4*(5), 1392–1410.
6. Lowe, D. G. (1999). Object recognition from local scale-invariant features. In *The Proceedings of the Seventh IEEE International Conference on Computer Vision*, 1999 (Vol. 2). IEEE.
7. Bay, H., Tuytelaars, T., & Van Gool, L. (2006). Surf: Speeded up robust features. In *European Conference on Computer Vision*. Berlin: Springer.
8. Rublee, E., et al. (2011). ORB: An efficient alternative to SIFT or SURF. In *2011 IEEE International Conference on Computer Vision (ICCV)*. IEEE.
9. Retrieved from http://maps.google.com/

An Object Detection Algorithm for UAV Reconnaissance Image Based on Deep Convolution Network

Xinping Guo, Xiangbin Li, Qiufeng Pan, Peng Yue, and Jiaxing Wang

Abstract In recent years, the UAV technology has developed rapidly and played an important role in many fields, especially in intelligence, reconnaissance, and monitoring. Object detection can provide accurate target location and target category for reconnaissance missions, providing detailed command information for commanders. However, the current object detection algorithm based on deep convolution network does not work well on detection for small objects and so cannot be applied to small objects in the reconnaissance image of UAV. In this paper, an object detection algorithm for UAV reconnaissance image based on deep convolution network is proposed. The image is adaptively divided according to the UAV flight parameters and the payload parameters before sent into the network. Through this way, small objects can be located and classified in a high accuracy of location and classification. This method can detect objects with small size, multiple quantities, and multiple categories on UAV.

Keywords Deep learning · Convolution network · Object detection · Adaptive division · UAV

1 Introduction

Target detection is one of the three major tasks in the field of computer vision. It can locate and classify multiple objects in the image and has been widely used in many scenes. The process of object detection algorithm includes three steps: proposed region selection, feature extraction, and classification. By the way of extracting features with or without human designed features, object detection can be divided

X. Guo (✉) · X. Li · Q. Pan · P. Yue · J. Wang
Research Laboratory of Command and Control and System Simulation Technology, Beijing Aerospace Unmanned Vehicles System Engineering Research Institute, Beijing, China

E. T. Quinto et al. (eds.), *The Proceedings of the International Conference on Sensing and Imaging, 2018*, Lecture Notes in Electrical Engineering 606,
https://doi.org/10.1007/978-3-030-30825-4_5

into detection based on artificially designed features and detection based on deep learning automatic extracting features. The algorithm based on artificially designed features obtains proposed region by using a detection window to entirely search on different scales of the image, uses artificially designed features to extract features such as Haar [1], LBP [2], or HOG [3], and finally classifies into shallow classifier SVM [4] and decision tree [5]. These methods have achieved good results in the field of object detection based on artificially designed features but they do relatively poor on classification accuracy and speed. In recent years, with the rapid development of neural networks and big data technologies, more and more deep learning algorithms have been applied in the field of object detection and have made big breakthroughs. R Girshick et al. proposed a method by using Selective Search [6] to select proposed regions and RCNN (Regions with CNN features) [7] and Fast RCNN (Fast Regions with CNN features) [8] convolution network to extract features. The classification accuracy has been greatly improved. S Ren et al. proposed Faster RCNN (Faster Regions with CNN features) [9] and the detection speed is greatly improved without loss of classification accuracy, so that the computer detection capacity gradually approaches the human level. However, the deep learning method mentioned above has a relatively poor result on small objects, and when the UAV performs the reconnaissance task at a relative altitude of 2 km from ground, the objects captured by the payload on UAV are usually small, which limits the method to be used on UAV.

This paper proposes an object detection based on Faster RCNN network for UAV reconnaissance images. By establishing a flight image dataset, the Faster RCNN algorithm is trained and the algorithm can adaptively divide the image according to the payload parameters and the minimum detectable pixel size calculated which is decided by the algorithm and training set. As a result, a small object with approximately 1% of the whole image size can be detected, thereby achieving object detection with small sizes, multiple categories, and multiple numbers.

2 Faster RCNN Algorithm

Faster RCNN algorithm is one of the most popular deep learning networks used for object detection, and its classification accuracy and speed have reached the forefront of current detection algorithms. The algorithm is improved on the basis of RCNN and fast RCNN. The proposal regions' selection, feature extraction, classification, and location refinement are unified into a deep network framework as shown in Fig. 1, which improves the comprehensive performance.

Faster RCNN algorithm has better performance on precision and recall compared to other detection algorithms based on deep learning, just as shown in Table 1. It can be concluded from the table that Faster RCNN is suitable for reconnaissance image on UAV because of its better precision and recall on small targets.

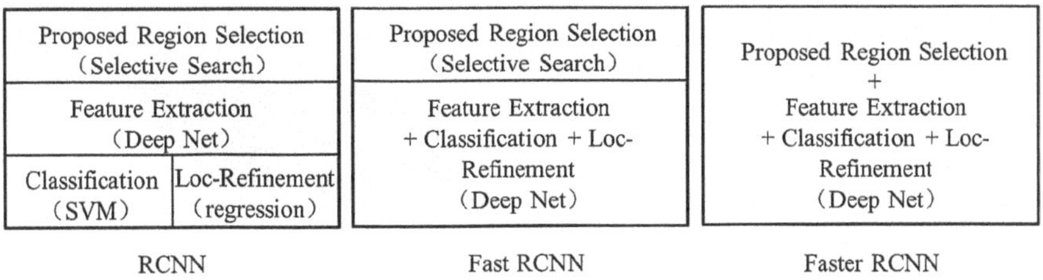

Fig. 1 The structures of three object detection frameworks

Table 1 Performance of four detection algorithms on COCO dataset [10]

Algorithm	AP (average precision)					AR (average recall)					
	$AP^{0.5}$	$AP^{0.75}$	S	M	L	1	10	100	S	M	L
Faster RCNN	**45.3**	**23.5**	**7.7**	**26.4**	37.1	**23.8**	**34.0**	34.6	**12.0**	**38.5**	54.4
Fast RCNN	39.9	19.4	4.1	20.0	35.8	21.3	29.5	30.1	7.3	32.1	52.0
SSD300	41.2	23.4	5.3	23.2	39.6	22.5	33.2	35.3	9.6	37.6	56.5
YOLOv2	44.0	19.2	5.0	22.4	35.5	20.7	31.6	33.3	9.8	36.5	54.4

Faster RCNN Network Architecture

Faster RCNN network has three parts: convolution and pooling network, region proposal network, and classification network as shown in Fig. 2.

The convolution and pooling network consists of a few of convolution layers and pooling layers, which is responsible for the feature extraction of the UAV flight image. The convolution layers have two dimensions: depth and width. Convolution kernels in width are responsible for detecting different features and kernels in depth are responsible for combining simple features from lower layers to complex features. After the convolution and pooling network, the UAV image is converted into a feature map, representing the key features of the objects.

The RPN consists of a convolution layers and a softmax layer. The network can generate the probabilities p_{obj} of the foreground and background of each anchor in each feature. The anchors are generated with different sizes and scales for each feature in the feature map. Meanwhile, the network can generate the deviation (pl_x, pl_y, pl_w, pl_h) of center position and sizes between this anchor and its true label window. Then the anchors are selected with the top-N probabilities as proposed regions to be sent into the subsequent network for training.

The classification network is composed of a fully connected layer and a softmax layer. The network uses the proposed regions generated by RPN, the feature map generated by convolution and pooling network, and the ground-truth labels as input to calculate the probability p_{cls}; if this region contains an object, then its position deviation is (l_x, l_y, l_w, l_h).

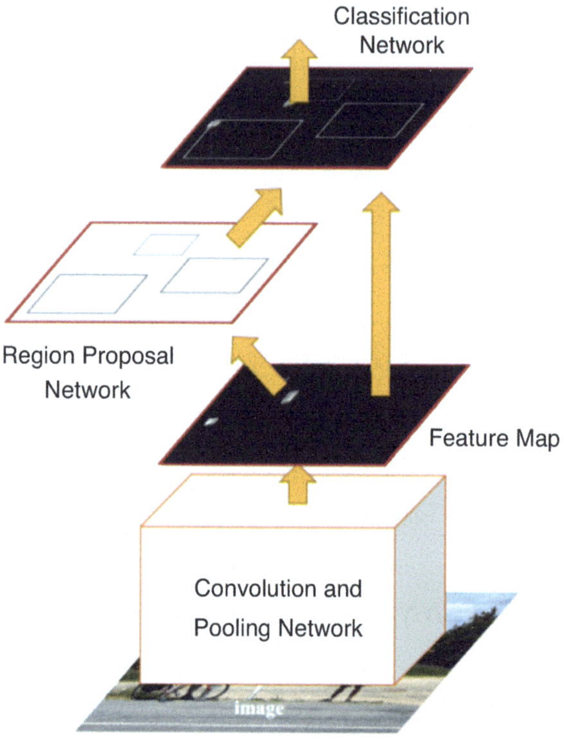

Fig. 2 Faster RCNN network architecture

Loss Function

The loss function can provide the target for training the network. The loss function of the Faster RCNN algorithm consists of two parts: the loss of RPN and the loss of classification.

RPN Loss Function

In order to train the RPN network, each anchors needs to be assigned a label {contains object, non-object} and the following conditions are used to define the positive label (contains object).

Positive label:
- The biggest overlap with ground-truth window
- Overlaps with ground-truth window which are bigger than 70% of total size

Therefore, the RPN loss function can be defined as Eq. (1),

$$L_{\text{RPN}} = \frac{1}{N_{\text{cls}}} \sum_i L_{\text{cls}} \left(p_i, p_i^* \right) + \lambda \frac{1}{N_{\text{reg}}} \sum_i p_i^* L_{\text{reg}} \left(t_i, t_i^* \right) \tag{1}$$

in which p_i represents the probability of the anchor containing an object, p_i^* represents the true label of each anchor, t_i represents the four parameterized coordinates of the predicted box, and t_i^* is the coordinates of the ground-truth window with a positive label. $L_{\text{cls}} \left(p_i, p_i^* \right)$ is the foreground and background classification loss of each anchor, which is shown as Eq. (2)

$$L_{\text{cls}} \left(p_i, p_i^* \right) = -p_i^* \log \left(p_i \right) - \left(1 - p_i^* \right) \log \left(1 - p_i \right) \tag{2}$$

$L_{\text{reg}} \left(t_i, t_i^* \right)$ is the regression loss which can be calculated by Eq. (3)

$$L_{\text{reg}} \left(t_i, t_i^* \right) = \begin{cases} 0.5 \left(t_i - t_i^* \right)^2, & \text{if } \left| t_i - t_i^* \right| < 1 \\ \left| t_i - t_i^* \right| - 0.5, & \text{otherwise} \end{cases} \tag{3}$$

N_{cls}, λ, and N_{reg} are the hyper-parameters to adjust the weights between these two losses.

Classification Loss Function

The formulation of the classification loss function L_{CLS} is similar to Eq. (1), except that the foreground and background classification loss $L_{\text{cls}} \left(p_i, p_i^* \right)$ is changed to multi-class classification loss shown as Eq. (4),

$$L_{\text{cls}} \left(p_i, p_i^* \right) = - \sum_i^C p_i^* \log \left(p_i \right) \tag{4}$$

where C is the number of all categories.

Therefore, the total loss function of the Faster RCNN algorithm is

$$L_{\text{total}} \left(p_i, p_i^* \right) = L_{\text{RPN}} \left(p_i, p_i^* \right) + L_{\text{CLS}} \left(p_i, p_i^* \right) \tag{5}$$

3 Adaptive Image Division Algorithm

According to the payload azimuth, flight height, field size, and pixel size, adaptive image division algorithm automatically divides the image into several blocks and resizes the blocks to the original size of image to satisfy the minimum detectable pixel size. Through this way, small objects can be detected in the image. The specific process is as follows:

1. Determine the minimum detectable pixel size (MDPZ) to be detected of each category, and take car class as an example, MDPZ of the car is $a \times b$.
2. According to the actual detection size of object and the posture of the UAV and payload, calculate the pixel size $a_1 \times b_1$ that the object occupies in the image.
3. If $a_1 < a$ or $b_1 < b$, divide the image into $N \geq c/(c_1 \times \text{scale})$ blocks, and

$$c = \begin{cases} a, & if\ \frac{a}{a_1} \geq \frac{b}{b_1} \\ b, & \text{else} \end{cases}$$

   scale is the resize scale in the preprocessing progress.
4. Resize each of the blocks to the original size and send to the network.
5. Finally, integrate the detection results of each block.

4 Experiments and Results

Throughout the experiment includes five parts: establishing the training set, labeling the objects, training, validating, and testing. The flow chart is shown in Fig. 3.

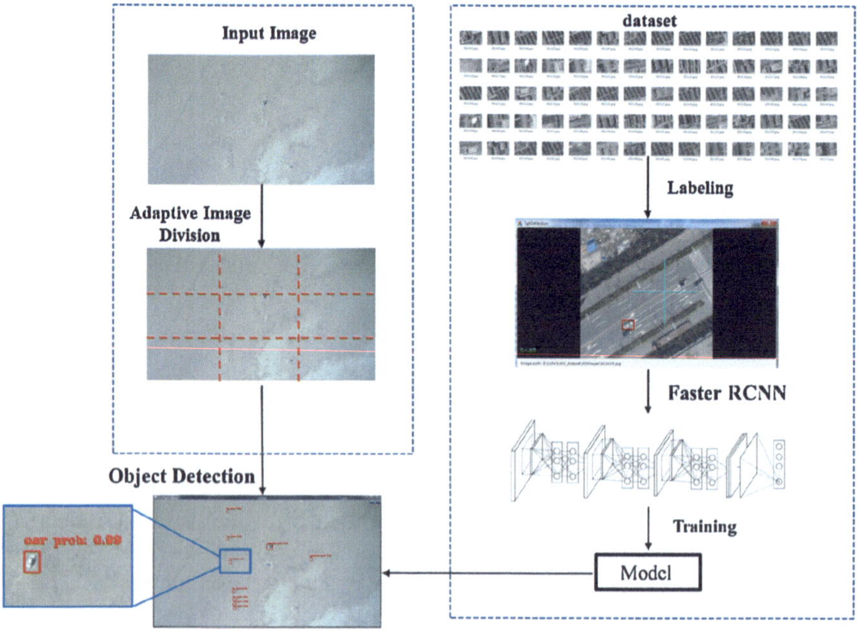

Fig. 3 Flow chart of the Faster RCNN network in the reconnaissance images of UAV

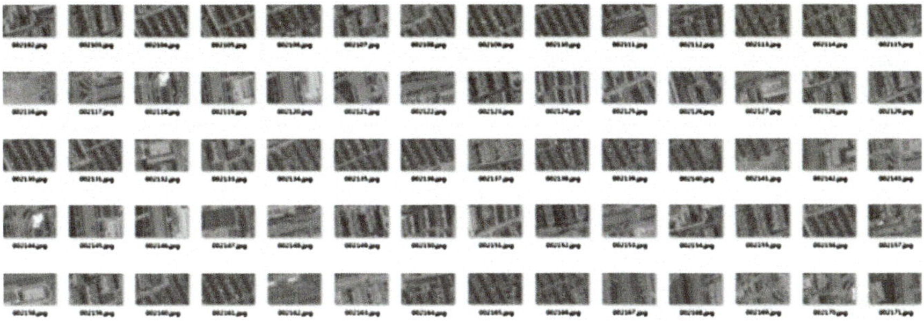

Fig. 4 Schematic diagram of part of the training set in the experiment

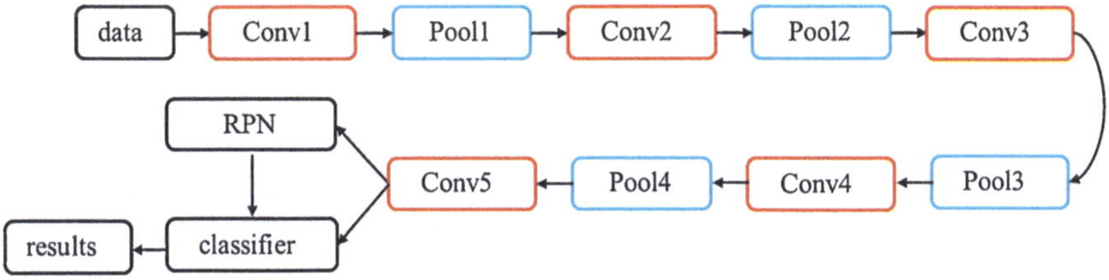

Fig. 5 VGG16 network structure in this experiment

Building Training Set and Labeling

The training set of this experiment contains about 4500 images recorded by visible light camera in which includes totally 10,000 targets (aeroplanes, cars, and boats), as shown in Fig. 4. The dataset is constructed by manually labeling, coordinate storage, and xml conversion.

Network Training

In this experiment, the convolution and pooling network uses the VGG16 network and it has five sets of convolution layers and pooling layers. The network architecture diagram of VGG16 is shown in Fig. 5.

The training set is sent into the network for training, using the stochastic gradient descent as optimizing method. One image was sent into the network on each iteration. After 10,000 iterations training on the training set, the loss after training was less than 0.3 and the accuracy was about 85% on training set. The changes in training loss with training iterations are shown in Fig. 6.

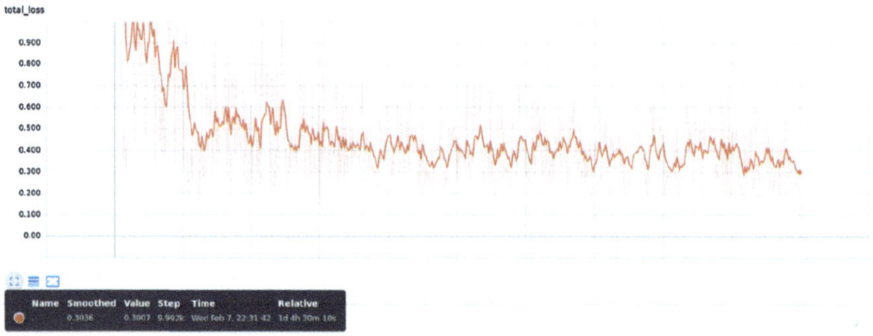

Fig. 6 Training loss decreases with the training iterations

Network Validation

The cross-validation method has strong stability and fidelity and can be applied for different sizes of training sets. Therefore, k folds cross-validation method is used in this experiment to validate the performance of the model. This method divides the dataset D into k mutually subsets with the same size $D = D_1 \cup D_2 \cup \ldots \cup D_k$, and each subset is sampled from D in layers. Take every $k - 1$ subsets as the training set and the reminder subset as the validation set, then repeat this process k times and get the average validation result of these k results. The entire process is shown in Fig. 7. The trained model was validated on the validation set of UAV reconnaissance images using cross-validation method and the mean loss was around 0.3, as shown in Fig. 8.

Results on Reconnaissance Images

The trained model was tested on the test set which contains aeroplane, boat, and car. The AP (Average Precision) for each type and mAP (mean Average Precision) is shown in Table 2. The mAP can achieve around 0.79 and is slightly improved by AID. The recall has a greatly improved by using AID.

The trained model is tested to measure the minimum detectable pixel size and the results are shown in Table 3. The results can be the condition of the adaptive division.

The trained model is used on the UAV scout images as shown in Fig. 9, and it can be seen that the algorithm can handle with small size, multiple categories, and multiple numbers.

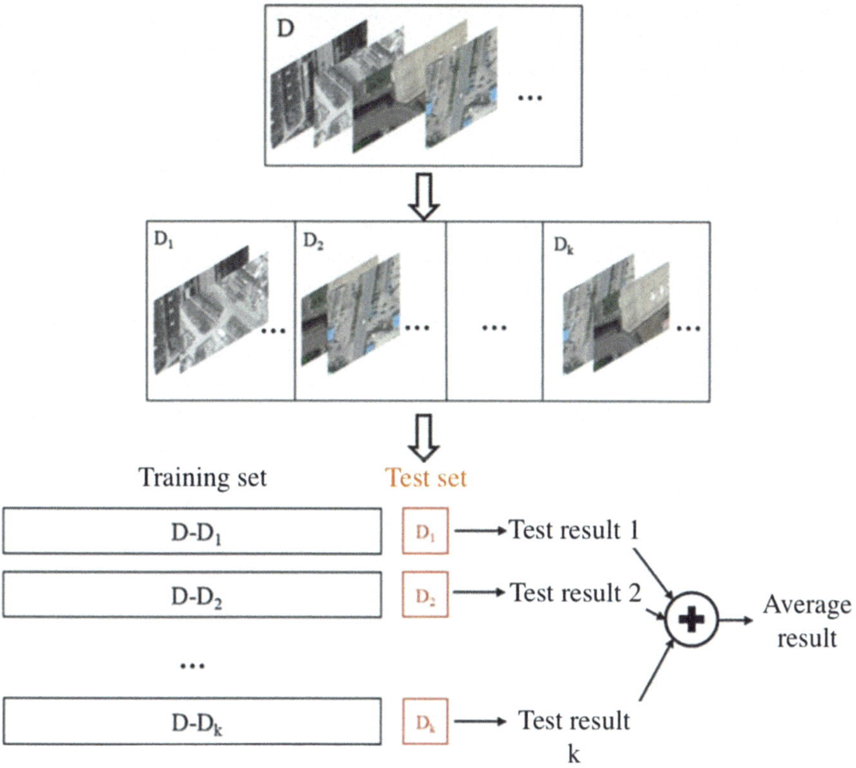

Fig. 7 Cross-validation method

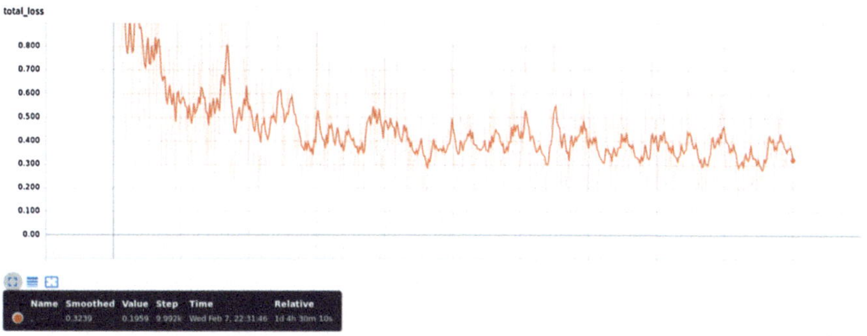

Fig. 8 Loss on validation set changes with training iterations

Table 2 Average precision on objects of different types

Type	Aeroplane	Boat	Car	mAP	Recall
Without AID	**0.94**	**0.73**	**0.70**	**0.79**	**0.31**
With AID	**0.95**	**0.77**	**0.76**	**0.83**	**0.74**

Table 3 Minimum detectable pixel size for different categories

Type	Aeroplane	Boat	Car
Minimum pixel size that can be detected (width × height)	28 × 30	18 × 24	20 × 25

5 Conclusion

This paper analyzes the characteristics of the objects in UAV reconnaissance images and introduces a deep convolution network algorithm suitable for objects with small sizes, multiple categories, and multiple quantities. By building the dataset, training the network on the dataset and dividing the image according to the minimum detectable pixel size, the algorithm for UAV reconnaissance images is implemented and can get a good result on validation set. The algorithm can achieve better results by expanding the numbers and the categories of the objects in training set.

Fig. 9 The results of the algorithm in UAV scout images. (**a**) Object's size is 1% of the image size. (**b**) Number of objects in one image is larger than 30. (**c**) Images contain many objects with multiple categories

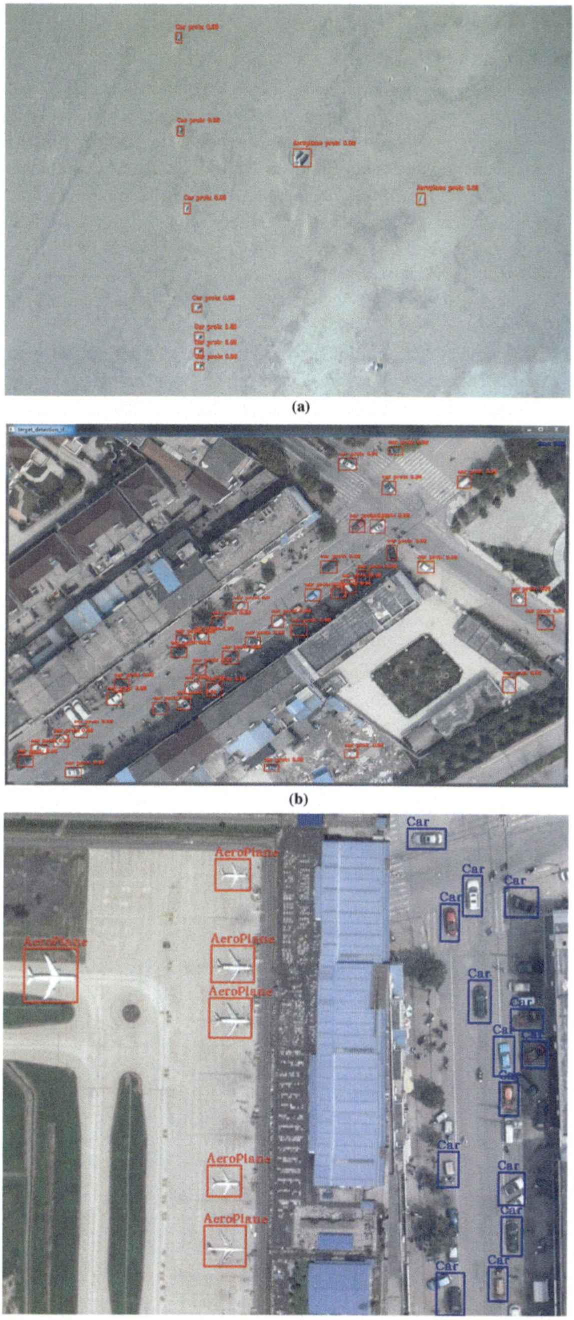

References

1. Bay, H., Ess, A., Tuytelaars, T., & Van Gool, L. (2008). Speeded-up robust features (SURF). *Computer Vision and Image Understanding, 110*(3), 346–359.
2. Ojala, T., Pietikainen, M., & Maenpaa, T. (2000). Multiresolution gray-scale and rotation invariant texture classification with local binary patterns. *IEEE Transactions on Pattern Analysis & Machine Intelligence, 24*(7), 971–987.
3. Dalal, N., & Triggs, B. (2005). Histograms of oriented gradients for human detection. In *IEEE Computer Society Conference on Computer Vision & Pattern Recognition* (pp. 886–893). IEEE Computer Society.
4. Suykens, J. A. K., & Vandewalle, J. (1999). Least squares support vector machine classifiers. *Neural Processing Letters, 9*(3), 293–300.
5. Safavian, S. R., & Landgrebe, D. (2002). A survey of decision tree classifier methodology. *IEEE Transactions on Systems Man & Cybernetics, 21*(3), 660–674.
6. Uijlings, J. R. R., Sande, V. D., et al. (2013). Selective search for object recognition. *International Journal of Computer Vision, 104*(2), 154–171.
7. Girshick, R., Donahue, J., Darrell, T., & Malik, J. (2014). Rich feature hierarchies for accurate object detection and semantic segmentation. In *Proceedings of the IEEE Conference on Computer Vision and Pattern Recognition* (pp. 580–587).
8. Girshick, R. (2015). Fast R-CNN. In *IEEE International Conference on Computer Vision* (pp. 1440–1448). IEEE Computer Society.
9. Ren, S., He, K., & Girshick, R., et al. (2015). Faster R-CNN: Towards real-time object detection with region proposal networks. In *International Conference on Neural Information Processing Systems* (pp. 91–99). MIT Press.
10. Liu, W., Anguelov, D., Erhan, D., Szegedy, C., & Reed, S. E. (2015). *SSD: Single shot multibox detector*. CoRR, abs/1512.02325.

Visualization of Horizontal Settling Slurry Flow Using Electrical Resistance Tomography

Kun Li, Mi Wang, and Yan Han

Abstract Settling slurry flow is very common and important in many industries, especially in transportation, which need to be monitored in practical operation. An investigation on visualization of horizontal settling slurry flow in pipeline using electrical resistance tomography was made in this paper. The internal images of fluid structure were displayed to operators with measurement of the solids concentration distribution and solids velocity distribution in pipe cross section. Experimental investigation with 5% solids loading concentration at various transport velocities was conducted. Meanwhile, the results of photography and other flow measurement methods were compared with the results obtained from electrical resistance tomography.

Keywords Flow visualization · Horizontal settling slurry flow · Electrical resistance tomography · Local solid concentration · Local solid velocity

1 Introduction

The transportation of settling slurry flow is required in a diverse range of industries, such as mining, nuclear, energy, pharmaceutical, chemical, and food industries [1–4]; especially in some specific applications, hydraulic transport through pipelines is the only method to transport solid particles. As the slurry is an essential mixture of solid and liquid, its characteristics rely on many factors, such as size and orientation of pipes, size and concentration of solids, velocity and viscosity of the liquid carrier. Slurry transportation is a highly complex process [2], which should be under measuring and monitoring in the transport process, and appropriate

K. Li (✉) · Y. Han
Shanxi Key Laboratory of Signal Capturing and Processing, North University of China, Taiyuan, China

M. Wang
School of Chemical and Process Engineering, University of Leeds, Leeds, UK

E. T. Quinto et al. (eds.), *The Proceedings of the International Conference on Sensing and Imaging, 2018*, Lecture Notes in Electrical Engineering 606,
https://doi.org/10.1007/978-3-030-30825-4_6

slurry flow parameters can efficiently avoid pipe blockage, equipment failures, and environmental damage. With well understanding the internal structure of the slurry flow, the optimal design, accurate analysis, and safe operation of slurry transportation systems can be more easily achieved.

Due to the complex nature of slurry flow and the effect of gravity, it is very difficult for measuring and visualizing settling slurry flow. In the past, venturimeter [5] based on differential pressure technique was widely used for measuring fluid flow rate. Conductivity probes [6] were installed inside pipes to measure the solids volumetric concentration, which disturbs the internal structure of slurry flow (i.e., intrusive method), and the abrasive nature of slurry makes the probe sensor to hardly survive. With significant efforts of worldwide researchers, several non-intrusive methods [7–10] appeared, such as, optical (laser), nuclear (X-ray or gamma ray), ultrasound, and conductance transducers. Among the above methods, electrical resistance tomography (ERT) offered a good solution for measuring and visualizing settling slurry flow, since the optical method cannot measure muddy fluid, and nuclear method is expensive and radioactive.

Phase volume fraction and phase velocity are two important parameters to describe the internal structure of slurry flow, and flow characteristics also can be extracted from them, for example, flow regimes [3]. With ERT online measurement, the real-time solids concentration and velocity distribution are offered, which allows process owners to "see" inside the pipe and determine flow conditions. The information can be used for understanding and managing slurry flow, and they also provide an experimental basis for CFD and other models in complex flows. Therefore, this paper focuses on the visualization of horizontal settling slurry flow, where local solids volumetric concentration and local solids axial velocity are measured with using an ERT system.

2 Horizontal Settling Slurry Flow Regimes

A flow regime describes the solids spatial distribution of settling slurry, which is crucial for design, optimisation, and the control of slurry flow processes. In horizontal settling slurry flow, with the influence of gravity and various flow velocities, the separation of phases occurs and four main flow regimes (from high velocity to low velocity) are developed, namely pseudo-homogeneous, heterogeneous, moving bed, and stationary bed [2], as shown in Fig. 1.

Homogeneous flow regime usually occurs at high velocities, where solids are fine particles and fully suspended in the liquid carrier, as shown in Fig. 1a. As the solid particles are almost equally distributed and flow with same velocity from top to bottom of pipe, it can be approximately regarded as single phase flow, which allows the equivalent fluid model used for representing the type of flow. This flow regime is the mostly used one in industrial applications. As the flow velocity decreases, heterogeneous flow regime occurs, where the solids concentration gradient and axis solid velocity gradient appears at the cross section of pipe. As shown in Fig. 1b, the

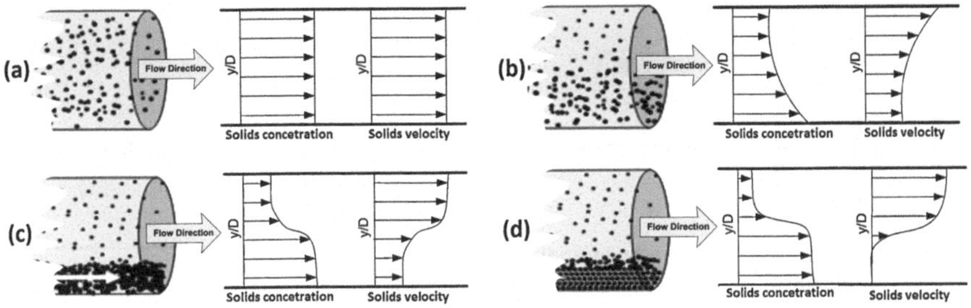

Fig. 1 Schematic presentations of horizontal slurry flow regimes with solids concentration and axis velocity profile. (**a**) Pseudo-homogeneous flow regime. (**b**) Heterogeneous flow regime. (**c**) Moving bed flow regime. (**d**) Stationary bed flow regime

flow regime is referred to as an intermediate flow regime, or might be in the process of forming a flowing bed, where some finer particles are suspended at top part of pipe and coarser particles are suspended at bottom part of pipe. The flow regime is usually applied in mining and dredging applications but a critical velocity is needed to maintain the flow regime, i.e., minimum velocity of liquid carrier. Moving bed flow regime starts to be formed when the flow velocity is below the critical velocity, where larger particles accumulate at the bottom of pipe and form a flowing bed, and the upper part of fluid is still heterogeneous mixture with less solids concentration, as shown in Fig. 1c. The solids concentration at bottom is maximum and decreases from bottom to top at cross section of pipe, while the solids velocity at upper part is higher than that at bottom part. As the flow velocity reduces further, the liquid carrier cannot move the solids on bed, which will be stationary and contact with the bottom of pipe, as shown in Fig. 1d. With the accumulation of solid particles, the stationary bed turns thicker, which might lead to blockage. Therefore, it is impossible for maintaining stationary bed flow regime for very long time, and it should be avoided in practice.

3 ERT System and Principle

The principle of ERT [11] is based on the concept of Ohm's law: by injecting a certain signal through electrodes pair into a conductive sensing region, and the injected signal will result in boundary voltages on the remaining electrodes. The boundary voltages contain the information of conductivity distribution which represents the phase distribution within the sensing region. Therefore, the boundary voltages data were collected and used to image the internal structure of multi-phase flow with using an image reconstruction algorithm [12].

A typical ERT system [11] is made of ERT sensor, data acquisition system, and image reconstruction system, as shown in Fig. 2. Each plane of ERT sensor is a set of equally spaced electrodes mounted around the pipe wall, which is in contact

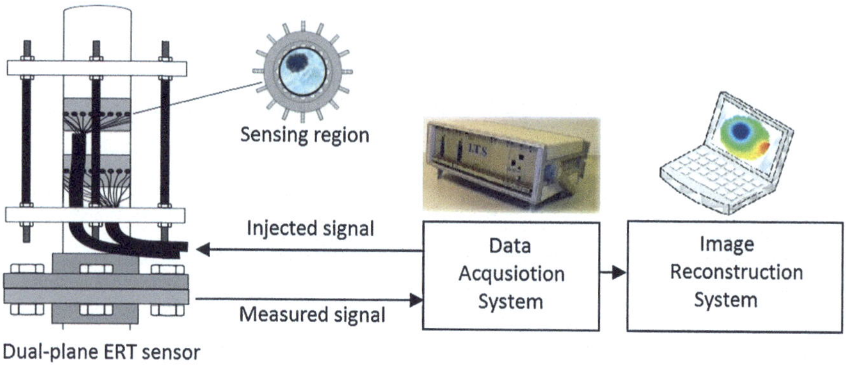

Fig. 2 Structure of ERT measurement system

with inside fluid, and does not cause any interference with the inside flow. The data acquisition system consists of signal sources, multiplexer arrays, voltmeters, signal demodulators, which is critical for ERT system since it determines sensing strategy, data accuracy, and highest data collection speed. Normally, an adjacent strategy [13] is used in most ERT systems, where a set of measured data contains 104 independent differential voltages for each 16-eletrode sensor. With using a preloaded image reconstruction algorithm, each set of measured data were transferred to each frame of internal conductivity distribution, which represents the phase distribution in the pipe cross section. The finite element mesh of pipe cross section is shown in Fig. 3.

Based on simplified Maxwell's relationship for slurry flow [14] (liquid carrier is water, and solids are silica sand in this paper), the obtained conductivity distribution can be used to derive the solids concentration distribution, as following Eq. (1).

$$\alpha = \frac{2\sigma_{\mathrm{w}} - 2\sigma_{\mathrm{m}}}{\sigma_{\mathrm{w}} + 2\sigma_{\mathrm{m}}} \tag{1}$$

where α is the solids concentration in each pixel, σ_{w} is the conductivity of continuous water phase, and σ_{m} is measured mixture conductivity of each pixel.

According to the internal conductivity distribution of dual plane, the pixel to pixel cross-correlation technique [15] is used to measure the time difference of solid particles flowing through two sensor planes, as shown in Fig. 4. With a predefined distance between them, the solids axial velocity can be calculated with using Eq. (2).

$$v = \frac{L}{T} \tag{2}$$

As the horizontal settling slurry flow without stirring can be assumed as vertically axial-symmetric, the solids concentration and velocity profile (Profile represents the distribution at vertical positions of pipe cross section) can be extracted using the average values of each row, i.e., Eqs. (3) and (4). And the local mean solids

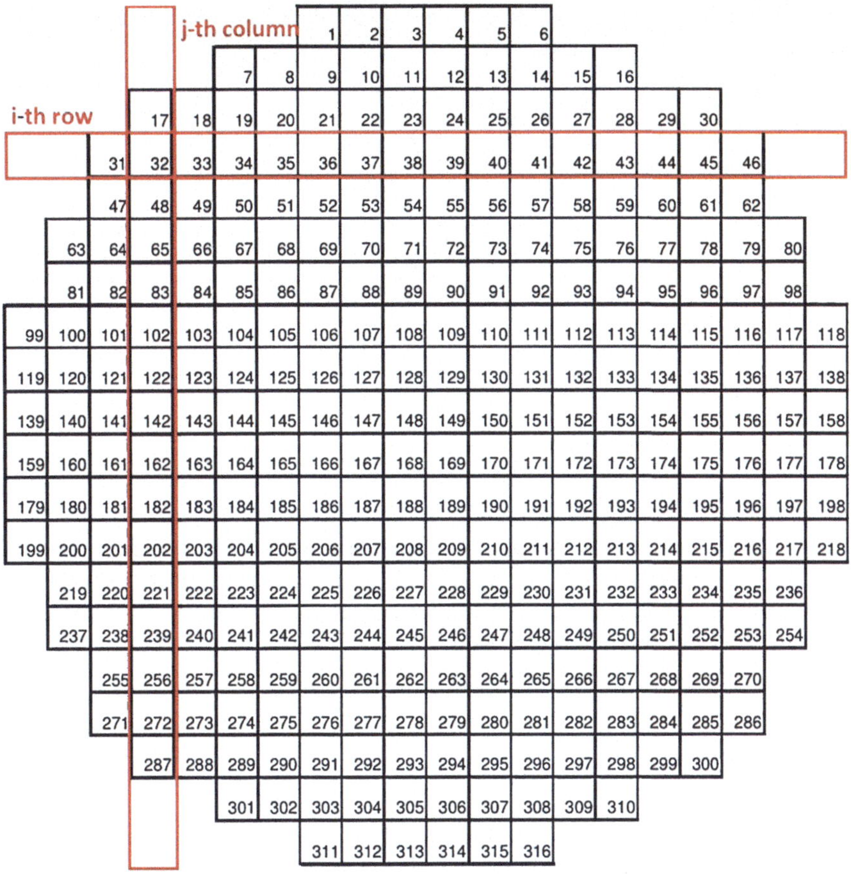

Fig. 3 Finite element mesh (*316 pixels*) of sensing region

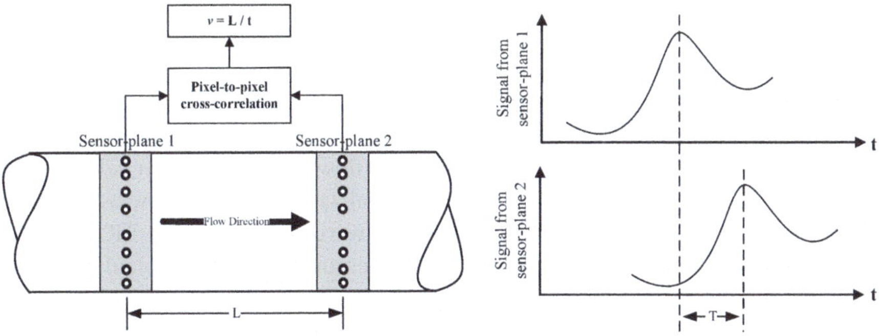

Fig. 4 Principle of pixel-to-pixel cross-correlation method

concentration and mean solids velocity can be calculated with the average value of 316 pixels, i.e., Eqs. (5) and (6).

$$\alpha_i = \frac{1}{n_i} \sum_{j=1}^{n_i} \alpha_{i,j} \tag{3}$$

$$v_i = \frac{1}{n_i} \sum_{j=1}^{n_i} v_{i,j} \tag{4}$$

$$\overline{\alpha} = \frac{1}{n} \sum \alpha_{i,j} \tag{5}$$

$$\overline{v} = \frac{1}{n} \sum v_{i,j} \tag{6}$$

where α_i and v_i are the average solids concentration and velocity of i-th row ($i = 1$, 2, ..., 20), respectively. $\overline{\alpha}$ and $\overline{v}$ are the local solids mean concentration and velocity, respectively. $\alpha_{i,j}$ and $v_{i,j}$ are the local solids concentration and velocity in the pixel of i-th row and j-th column, respectively.

4 Experiment Setup

Experimental investigation of settling slurry flow was conducted on a slurry flow loop facility in University of Leeds, as shown in Fig. 5. It consists of a high-performance dual-plane ERT system (FICA, built by OLIL) [16] for measuring local solids concentration and velocity, an EMF (OPTIFLUX 4300, from KROHNE) for obtaining fluid velocity, and a CMF (OPTIMASS 700 T50, from KROHNE) for obtaining slurry mass flow rate. The slurry flow run in experiment with 5% loading solids volumetric concentration at different transport velocity (from 1.5 to 4.0 m/s), where solid phase is silica sand (particle size is between 75 and 700 in diameter) and liquid carrier is tap water. Meanwhile, a section of transparent pipe was installed on slurry flow loop for taking photography of slurry flow at each transport velocity.

In order to evaluate the measurement and visualization of slurry flow using ERT, the photography results, EMF and CMF results were compared with ERT results. The solids concentration from CMF and EMF measurement can be calculated with Eq. (7).

$$\alpha \cdot \rho_{\mathrm{s}} + (1 - \alpha) \cdot \rho_{\mathrm{w}} = \frac{Q}{v \cdot A} \tag{7}$$

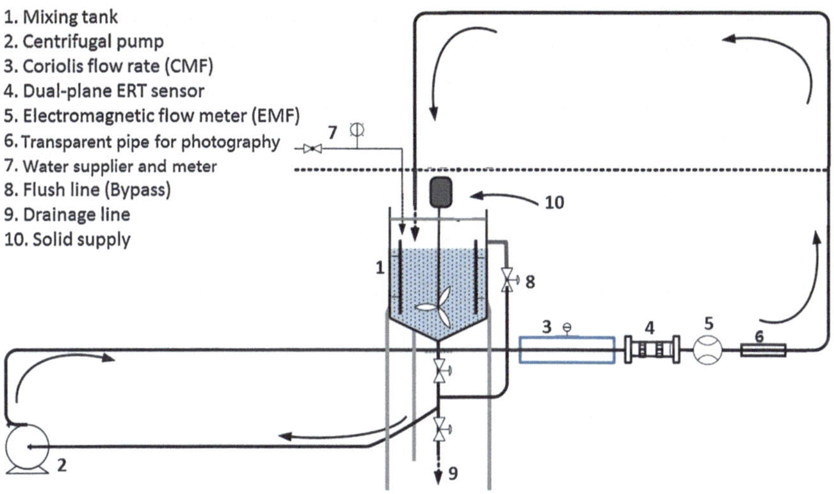

Fig. 5 Experimental facility (*slurry flow loop*) in Leeds University

where ρ_s and ρ_w are the density of sand and water, respectively. α is the solids concentration. Q and v are slurry mass flow rate and velocity, respectively. A is the area of pipe cross section.

5 Results and Discussions

Figure 6 shows the local solids concentration and solids velocity profile with different transport velocities. It can be noticed that the solids distribution and flow conditions depend on the transport velocity. At higher transport velocity, the solid particles are all suspended in slurry mixture, and almost uniformly distributed in the pipe cross section, as demonstrated at 4.0 m/s in Fig. 7a, and the solids velocities are almost equal in the pipe cross section. With the transport velocity decreasing, the solid particles in upper part tend to the lower part, and the velocity of solids particles in lower part slowdown, which tends to form a moving bed. Especially at 1.5 m/s, the solid particles clearly accumulate and move in lower part, as demonstrated in Fig. 7b, and the solids velocity in moving bed is lower than upper part.

Figure 8a shows that the local mean solids concentration obtained from ERT and from EMF and CMF are almost same (the measurement error is less than 1%), which implies that the ERT provides a reasonable method to visualize slurry flow. Figure 8b shows that the mean solids velocity obtained from ERT basically keep consistent with the transport velocity at high velocity (especially over 3.0 m/s). However, as the increase of solids concentration in pipe bottom leads to a strong

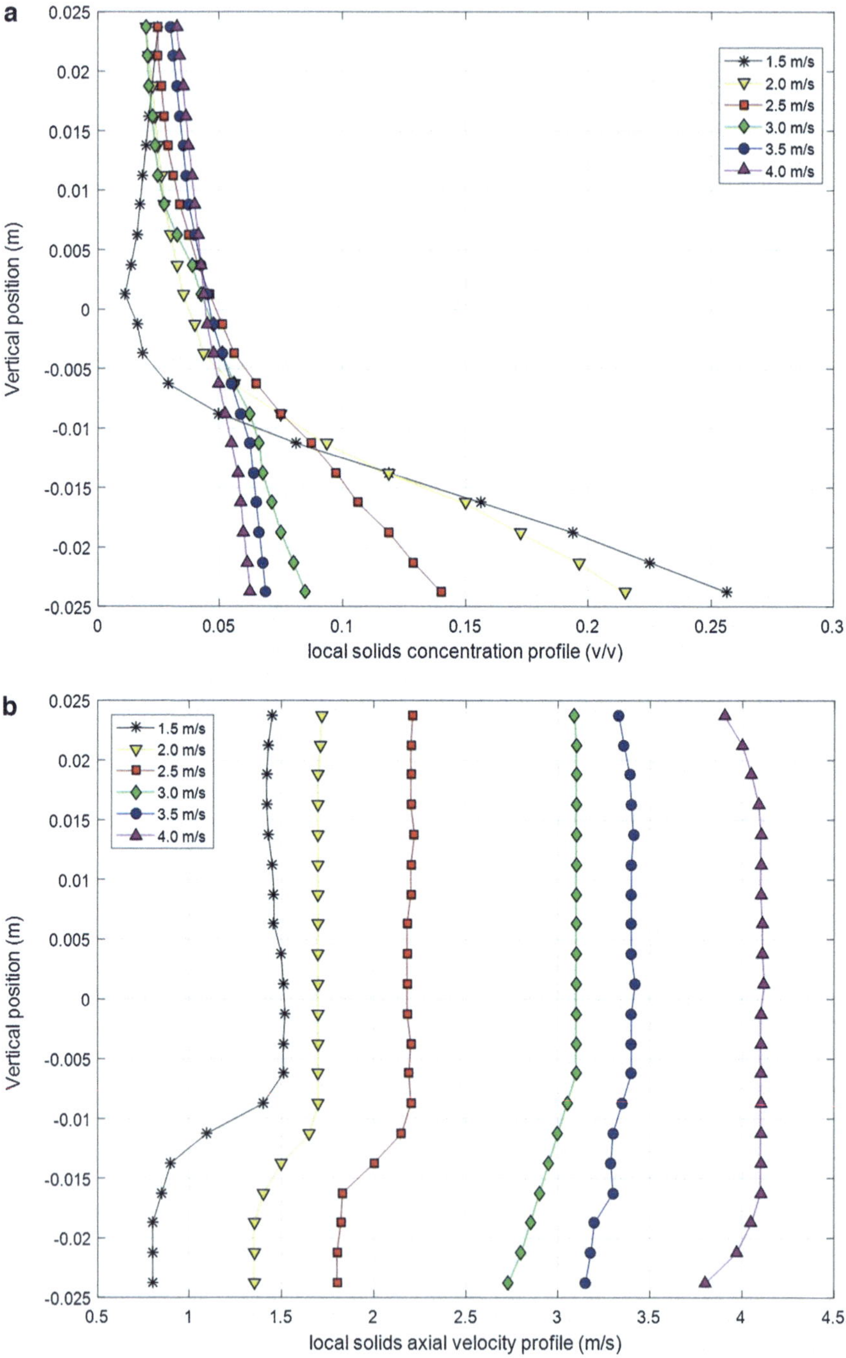

Fig. 6 Visualization of slurry flow with 5% sand (*transport velocities (from 1.5 to 4.0 m/s)*) (**a**) Local solids concentration profile. (**b**) Local solids velocity profile

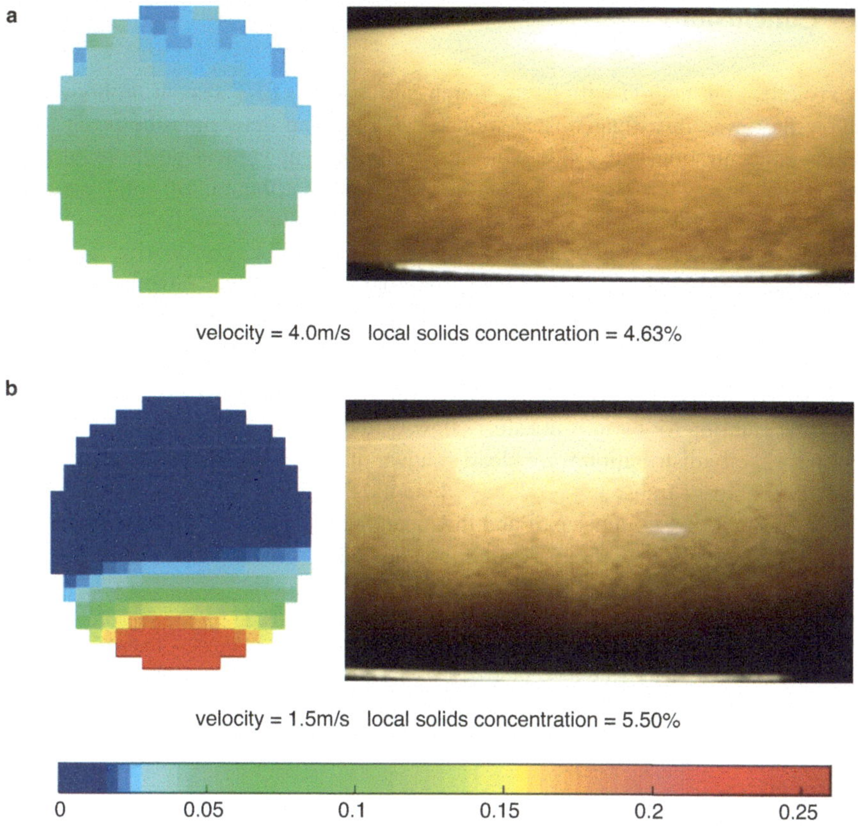

Fig. 7 Local solids concentration distribution of pipe cross section and photography of slurry flow. (**a**) Slurry flow at 4.0 m/s transport velocity. (**b**) Slurry flow at 1.5 m/s transport velocity

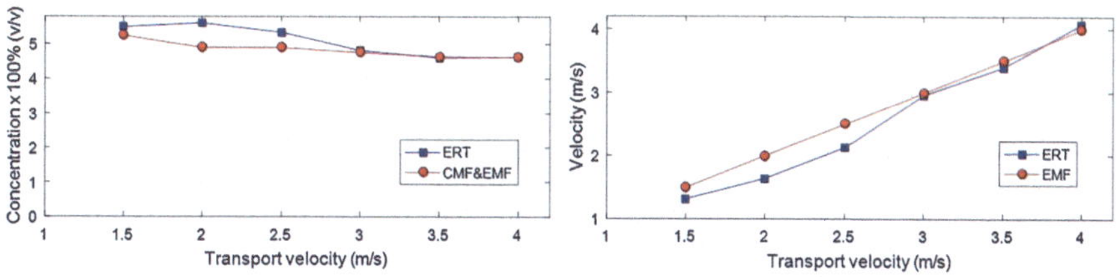

Fig. 8 Error analysis. (**a**) Comparison of local solids concentration obtained from ERT with that of CMF and EMF, (**b**) Comparison of solids velocity obtained from ERT and transport velocity measured by EMF

particle–particle interaction at low velocity (below than 2.5 m/s), the movement of solid particles is impeded, and the solids velocity is lower than transport velocity, which is highlighted by ERT.

6 Conclusions

A visualization method with using a high-performance ERT system on horizontal slurry flow was investigated. It demonstrated ERT can determine the mean local solids concentration and solids concentration profile at vertical positions of pipe cross section, which shows the amount and the distribution of solid particles in slurry. And with cross-correlation technique, the dual-plane ERT system can determine the mean local solids velocity and solids velocity profile in pipe. The ERT results were compared with actual photography and other flow measurement methods, which verified the dual-plane ERT system can perform well for visualizing slurry flow in pipeline. Meanwhile, the below conclusions were drawn:

- Compared with photography and other flow measurement methods, ERT offers a better solution for monitoring slurry flow, as the solids concentration distribution and velocity distribution were clearly shown in ERT results.
- At slow slurry transparent velocity, the solids velocity in lower part of pipe is slower than that in upper part and the mean local solids velocity is smaller than transparent velocity, which are highlighted in ERT results.

Acknowledgments The authors would like to express their gratitude for the support from the Chinese Scholarship Council (CSC) and the School of Chemical and Process Engineering, who made Mr. Li's study at the University of Leeds possible.

References

1. Stanley, S. J., & Bolton, G. T. (2010). A review of recent electrical resistance tomography (ERT) applications for wet particulate processing. *Particle & Particle Systems Characterization, 25*(3), 207–215.
2. Abulnaga, B. (2002). *Slurry systems handbook*. New York: McGraw-Hill. ISBN 0-07-137508-2.
3. Robert, C., & Ramsdell. (2013). An overview of flow regimes describing slurry transport. In *Proceedings of the Art of Dredging, Brussels, Belgium*.
4. Kaminoyama, M. (2014). Review of visualization of flow and dispersion states of slurry system fluids in stirred vessels. *Journal of Chemical Engineering of Japan, 47*(2), 109–114.
5. Shook, C. A., & Masliyah, J. H. (2010). Flow of a slurry through a venturi meter. *Canadian Journal of Chemical Engineering, 52*(2), 228–233.
6. Nasr-El-Din, H., Shook, C. A., & Colwell, J. (1987). A conductivity probe for measuring local concentrations in slurry systems. *International Journal of Multiphase Flow, 13*(3), 365–378.
7. Schmidt, M., Münstedt, H., Svec, M., Roosen, A., Betz, T., & Koppe, F. (2010). Local flow behavior of ceramic slurries in tape casting, as investigated by laser-Doppler velocimetry. *Journal of the American Ceramic Society, 85*(2), 314–320.
8. Shimada, T., Habu, H., Seike, Y., Ooya, S., Miyachi, H., & Ishikawa, M. (2007). X-ray visualization measurement of slurry flow in solid propellant casting. *Flow Measurement & Instrumentation, 18*(5), 235–240.
9. Krupička, J., & Matoušek, V. (2014). Gamma-ray-based measurement of concentration distribution in pipe flow of settling slurry: Vertical profiles and tomographic maps. *Journal of Hydrology & Hydromechanics, 62*(2), 126–132.

10. Stener, J. F., Carlson, J. E., Sand, A., & Pålsson, B. I. (2016). Monitoring mineral slurry flow using pulse-echo ultrasound. *Flow Measurement & Instrumentation, 50*, 135–146.
11. Wang, M. (2015). *Industrial tomography: Systems and applications*. Woodhead Publishing, Limited. ISBN 978-1-78242-118-4.
12. Li, Y., Cao, S., Man, Z., & Chi, H. (2011). Image reconstruction algorithm for electrical capacitance tomography. *Information Technology Journal, 10*(8), 269–291.
13. Wang, M., Wang, Q., & Bishal, K. (2016). Arts of electrical impedance tomographic sensing. *Philosophical Transactions, 374*(2070), 20150329.
14. Wang, M., Jones, T. F., & Williams, R. A. (2003). Visualization of asymmetric solids distribution in horizontal swirling flows using electrical resistance tomography. *Chemical Engineering Research & Design, 81*(8), 854–861.
15. MOSAIC Scientific Ltd. (2009). *AIMFlow Standard Version 1.0 user manual*. MOSAIC Scientific Ltd.
16. Wang, M., Ma, Y., Holliday, N., Dai, Y., Williams, R. A., & Lucas, G. (2005). A high-performance EIT system. *IEEE Sensors Journal, 5*(2), 289–299.

CSAR Moving Target Detection with Logarithm Background Subtraction Based on Optimal Azimuth Aperture Analysis

Wenjie Shen, Yun Lin, Fei Teng, Yanping Wang, and Wen Hong

Abstract This paper focuses on moving target detection in single-channel circular-SAR (CSAR). The logarithm background subtraction algorithm described in this paper utilizes the overlapped subaperture logarithm image sequence to detect moving targets. It first models the background image with the input image sequence, and then uses the input images to subtract the background image to cancel the clutter. Finally, the moving target can be detected in a subtracted image sequence. However the detection performance depends on the azimuth aperture width (or otherwise the number of input images in one set of image sequence). Thus, the detection performance is analyzed with two measurements: the signal-to-clutter noise ratio (SCNR) improvement and the clutter cancellation ability. Based on the analysis, the proper azimuth aperture width to achieve the best detection performance could be obtained. The algorithm is validated by the GOTCHA-GMTI dataset.

1 Introduction

Ground moving target detection is an important research area in SAR application because of its all-weather capability. Current applications such as battlefield surveillance and city traffic monitoring depend on multichannel techniques such as displaced phase center array (DPCA) [1], space-time adaptive processing (STAP)

W. Shen (✉) · F. Teng · W. Hong
Key Laboratory of Technology in Geo-spatial Information Processing and Application system, Institute of Electronics Chinese Academy of Science (IECAS), University of Chinese Academy of Science (UCAS), Beijing, China
e-mail: whong@mail.ie.ac.cn

Y. Lin · Y. Wang
School of Electronic and Information Engineering, North China University of Technology, Beijing, China
e-mail: ylin@ncut.edu; ypwang@ncut.edu

E. T. Quinto et al. (eds.), *The Proceedings of the International Conference on Sensing and Imaging, 2018*, Lecture Notes in Electrical Engineering 606,
https://doi.org/10.1007/978-3-030-30825-4_7

[2], along-track interferometry (ATI) [3]. Compared with the multichannel system, the single channel system is still an important part of the current system. Thus, research into moving target detection for single-channel systems is essential.

However, the detection capability of single-channel SAR is limited, as pointed out in Chapman et al. [4]. Recent research reveals that the circular geometry may resolve such single-channel SAR detection problems, such as Poisson et al. and Shen et al. [5, 6]. Our team proposed a moving target detection algorithm at the 2017 IGARSS [6]. A modified version is described in this paper. Compared with Shen et al. [6], the algorithm integrates the log-ratio operator to achieve a better clutter cancellation ability. The radiometric adjustment and constant false alarm rate (CFAR) detector are also added to achieve better performance. However, the optimal detection performance depends on the azimuth aperture width, which is used to generate the subaperture image sequence. Previously, the preset azimuth aperture width was mainly acquired by testing. Therefore, in this paper, we present an analysis on acquiring the parameter for achieving optimal detection performance.

The rest of the paper is organized as follows. Section 2 introduces the processing chain of the logarithm background subtraction algorithm. Section 3 consists of the method, whereas Section 4 presents the algorithm performance analysis, which is composed of data and experiment parameters, the signal-to-clutter noise ratio (SCNR) and clutter cancellation analysis, and the optimal detection result.

2 Processing Chain

The subaperture image can be decomposed into a background image (containing clutter) and a foreground image (contain a moving target). The position of the moving target' varies in a subaperture image sequence. The algorithm modeled the background image first, rather than simply making the difference between the subaperture images. The first advantage of this step is that it can avoid the target signature being canceled because of the use of the adjacent images performing subtraction. Another advantage is that the target SCNR could be improved by subtraction processing, which is good for moving target detection.

The processing chain is shown in Fig. 1. The algorithm consists of the following parts:

(1) Segment the full aperture into small arcs of equal length.

(2) For each arc, using the same image grid to generate an overlap subaperture logarithm image (OSLI) sequence.

(3) Performing the radiometric adjustment process on each OSLI sequence to remove the antenna pattern.

(4) Apply a median filter along the azimuth time dimension to the OSLI sequence to obtain the corresponding background image.

(5) Generating foreground images by using original OSLI subtracts the background image.

(6) The target is detected on the foreground images by a CFAR detector.

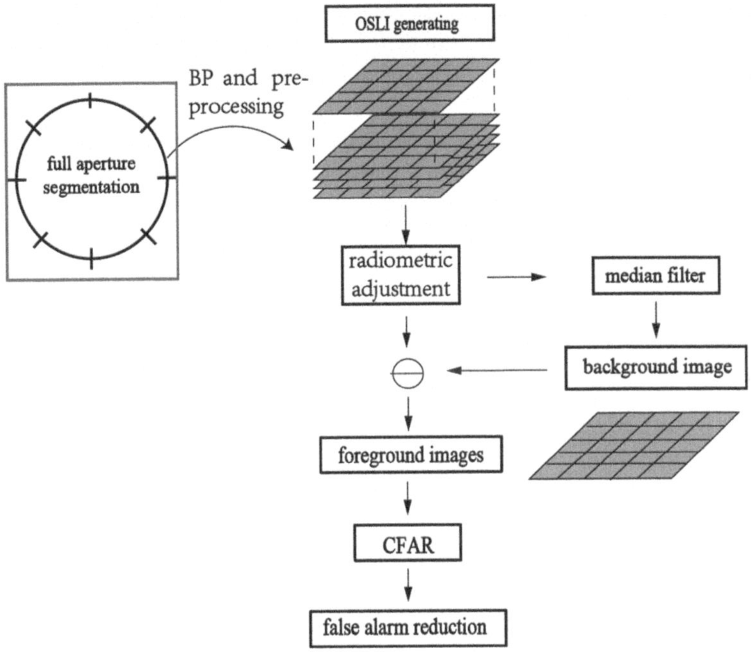

Fig. 1 Processing chain of the logarithm background subtraction algorithm

The moving target detection performance depends on the quality of the modeled background images. The quality of the modeled background images is controlled by the width of the input azimuth aperture (the arc mentioned above, otherwise the number of input images). Thus, in this paper, we mainly discuss the influence of the first step (labeled with a rectangle in Fig. 1) on detection performance.

3 Method

Detecting a moving target requires the SCNR to be high enough; thus, several techniques decrease the clutter energy first such as multichannel techniques. Logarithm background subtraction utilizes such an idea, but the difference is that the background is provided by using the overlapped subaperture image sequence. Thus, the background cancellation result influences the final detection performance. If the width of the input azimuth aperture is too narrow, there may be residual target signature in the background image. If the width of the input azimuth aperture is too wide, the background may mismatch with the clutter in the original image because of its anisotropic behavior. This causes a false alarm. Thus, it is natural to think that there is an optimal width of the input azimuth aperture value so that the target energy is well preserved whereas the clutter is mostly canceled. Two measurements, SCNR improvement and clutter cancellation quality, are used to find that value. The SCNR improvement factor is used to ensure that the target energy can be much preserved. Clutter cancellation quality is to avoid the defect caused by the anisotropic behavior of the clutter.

We find the proper width of the input azimuth aperture by carrying out the following step. First, we use the clutter cancellation quality data along the width of the input azimuth aperture to find the stable interval during which the clutter value does not vary too much. Then, we find maximum SCNR improvement within the interval. The corresponding angle is the optimal width of the input azimuth aperture.

Therefore, the proper width of the input azimuth aperture should meet the following condition: SCNR improvement reaches its maximum value during interval when the clutter is canceled most.

Improvement of SCNR is shown by the following:

$$G_{\mathrm{SCNR}}(\theta) = 10 \cdot \log \frac{\mu_t(\theta)}{\mu_c(\theta)} - 10 \cdot \log \frac{\mu_{t0}}{\mu_{c0}} \tag{1}$$

$G_{\mathrm{SCNR}}(\theta)$ is SCNR improvement. θ is the width of the input azimuth aperture. μ_{t0} and μ_{c0} are the maximum intensity of the target signature and the reference clutter in the original test frame. $\mu_{t0}(\theta)$ and $\mu_{c0}(\theta)$ are the intensity values extracted from the same position as μ_{t0} and μ_{c0} after subtracting the background generated with different θ. The clutter cancellation ability is evaluated by drawing the curve that $\mu_{c0}(\theta)$ changes along θ.

4 Algorithm Performance Analysis

In this section, we use a selected scene as an example to present the analysis on how to obtain the experiment parameters to achieve the optimal target detection performance with the logarithm background subtraction algorithm. For other CSAR data, the experiment parameter could be acquired by performing the same analysis.

Data and Experiment Parameters

To evaluate the performance, a test frame is selected as shown in Fig. 2. The SAR image is generated with channel one data in GOTCHA GMTI dataset [7]. The aperture width is 0.79°, the center time is 11.2 s. The image grid spacing is 0.2 m × 0.2 m. The target car Durango is delocalized owing to its motion, as labeled with a red rectangle. The reference clutter for SCNR improvement and clutter cancellation ability analysis is labeled with a blue circle.

To study the influence of input azimuth arc size on detection, the 20° azimuth aperture (the test frame is at 0°) is used to generate the overlapped subaperture

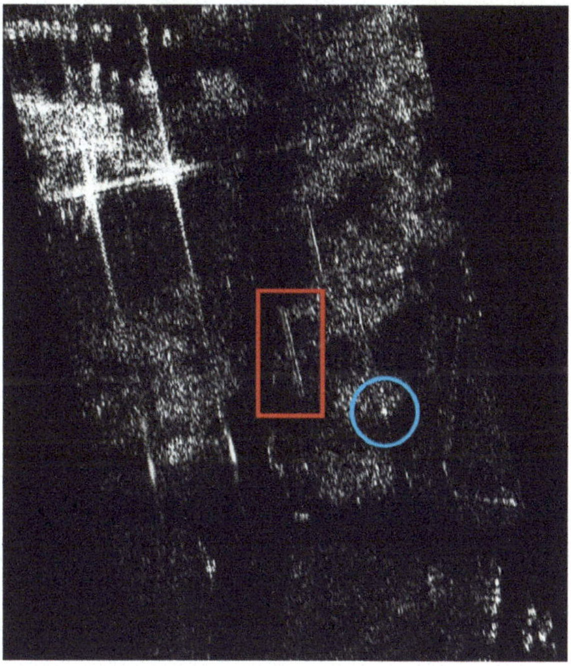

Fig. 2 Selected test frame. The target car Durango is delocalized owing to its motion as labeled with a red rectangle. Reference clutter for signal-to-clutter noise ratio (SCNR) improvement and clutter cancellation ability analysis is labeled with a blue circle

image sequence. Each subaperture image is generated using the same parameters as the test frame. The gap of the azimuth angle between adjacent images is 0.2°). Therefore, we have 100 images for evaluation. To calculate the SCNR, a strong stationary target, which exists within all 20° azimuth apertures, is selected as the reference. The clutter cancellation ability is also evaluated with the strong stationary target. After acquiring the parameter for optimal detection, the example detection result is shown on the test frame.

SCNR Improvement and Clutter Cancellation Analysis

According to the above measurements, we can find the proper θ, which fulfills the above condition as the experiment parameter for optimal detection.

The SCNR improvement curve is shown in Fig. 3.

In Fig. 3, the Durango's SCNR improvement factor increases fast along the θ at [0°, 6°] (the corresponding input images are from 0 to 30) and reaches the peak at 6°. This is because the quality of the background image is improved and the target signature is gradually filtered, as shown in Fig. 4.

In Fig. 4b, the target signature becomes weak but is still visible. In Fig. 4c, the target signature is almost filtered; thus, the improvement is quite high, at around 9 dB. In Fig. 4d, the target signature is fully filtered and has a good improvement

Fig. 3 Signal-to-clutter noise ratio improvement curve

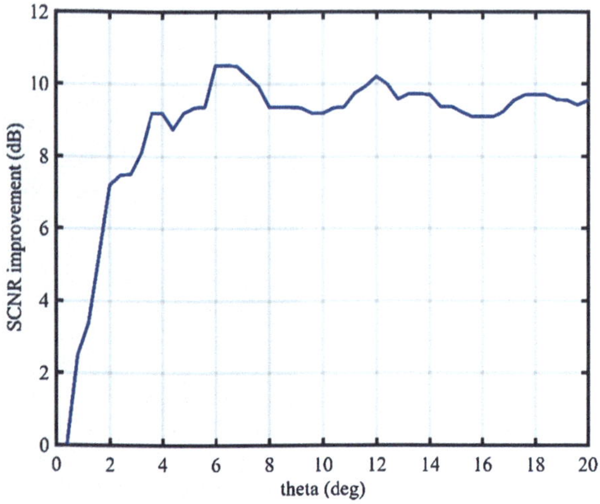

value at 10.52 dB. When $\theta > 6°$, the improvement factor decreases and remains stable at around 9–10 dB. This is because, although the target signature is fully filtered when $\theta > 6°$, the clutter value varies because of its anisotropic scattering behavior. Thus, the causes of the clutter in the test frame cannot be canceled by subtraction processing with model background images.

Take the reference clutter as the example shown in Fig. 5. As mentioned above, the reference clutter could be seen during all 20°. Therefore, we collect the residual clutter value from the same position as in the test frame after subtraction processing to evaluate the clutter cancellation ability. In Fig. 5, the residual clutter values are close to 0 dB during [0°, 12°]. This means that its scattering field is stable; therefore, the value is almost equal in the test frame and modeled background image. When $\theta > 12°$, the value is changed because of the anisotropic scattering behavior, thus causing the residual clutter value to gradually increase.

As mentioned previously, the experiment parameter for optimal target detection should fulfill two conditions. Thus, according to the above analysis, the proper width of the input azimuth aperture is $\theta = 6°$ (30 images). In the next section, we present the optimal detection result.

Optimal Moving Target Detection Result

The background image is obtained using 30 overlapped subaperture images. The corresponding background image is shown in Fig. 6.

Compared with test frame shown in Fig. 2, the moving target signatures have been filtered and the clutter structures such as the road edge and building are all preserved. Then, performing the subtraction step, the output image is shown in Fig. 7.

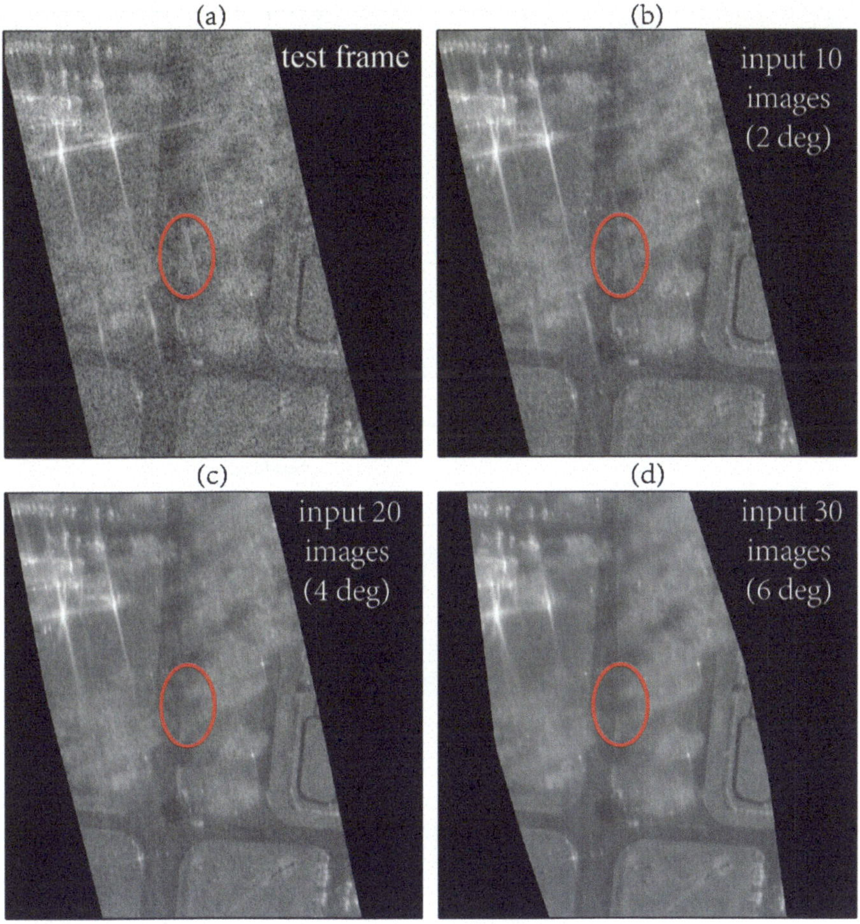

Fig. 4 Test frame and background image modeled with different input image numbers (azimuth aperture width). (**a**) Test frame. (**b–d**) Corresponds to the background generated with 10, 20, and 30 input images ($\theta = 2°; 4°; 6°$) respectively. The target signature position is labeled with a red circle

It could be seen in Fig. 7 that the moving target is well preserved. Except for the moving target signatures, the side lobe of the strong stationary target also exists. This is because the side lobe of this stationary target rotates during the 30 input images. Therefore, it is filtered in the background and the target's sidelobe is preserved in the subtracted image.

Apply a CFAR detector to a subtracted image and post-target discrimination processing to reduce the false alarm; the detection result is shown in Fig. 8. The Durango and the potential moving target are labelled in red. At the top left of the figure, the sidelobe of the two strong manmade structures is also detected. This is

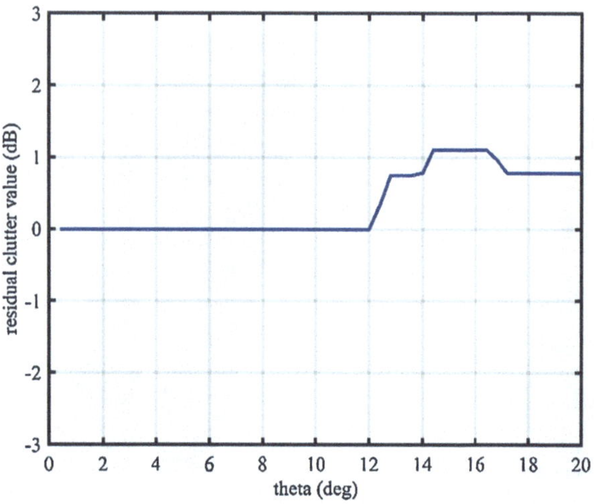

Fig. 5 Residual reference clutter curve

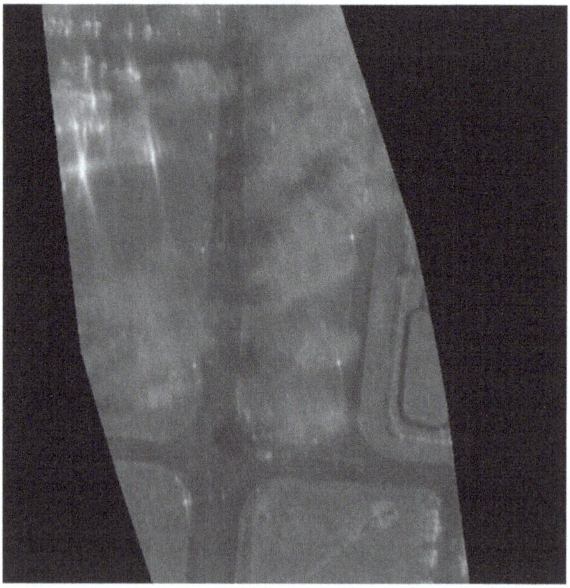

Fig. 6 Modeled background image

because the anisotropic target's rotating sidelobe is not preserved in the background image. It is not canceled; thus, it is falsely detected. The anisotropic behavior-induced false alarm may be excluded by better target discrimination processing in the future.

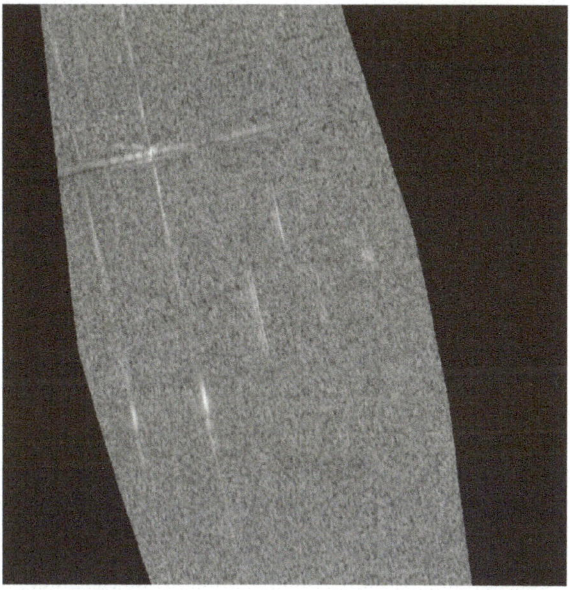

Fig. 7 Test frame after subtraction processing

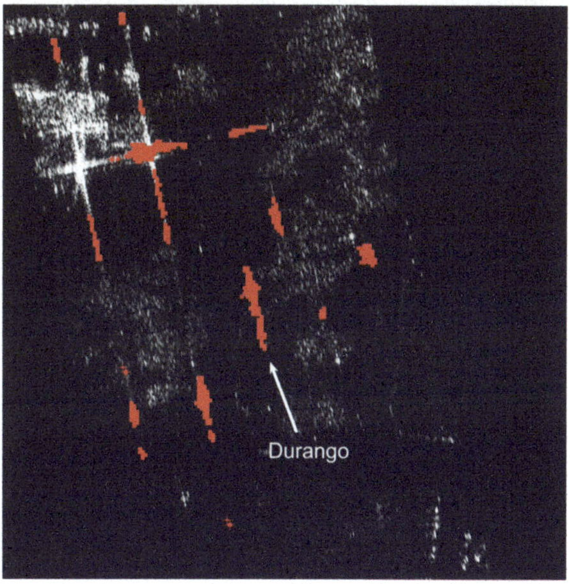

Fig. 8 Detection result

5 Conclusion

In this paper, the performance of the logarithm background subtraction algorithm is evaluated with SCNR improvement and clutter cancellation ability, two measurements. Following the presented analysis, the parameter of the proper number of input images (or otherwise the width of the input azimuth aperture) for achieving the optimal moving target detection can be determined. Other CSAR data that use the logarithm background subtraction algorithm can follow the same routine. Future work will involve studying better target discrimination processing to reduce the false alarm induced by the anisotropic backscattering behavior of the clutter.

Acknowledgments This work was supported by the Scientific Research Foundation of North China University of Technology, and supported by the Science and Technology Innovation Project of North China University of Technology. This work was also supported by the National Natural Science Foundation of China under Grant 61571421 and Grant 61431018.

References

1. Cerutti-Maori, D., & Sikaneta, I. (2013). A generalization of DPCA processing for multichannel SAR/GMTI radars. *IEEE Transactions on Geoscience and Remote Sensing, 51*, 560–572.
2. Ender, J. H. G. (1999). Space-time processing for multichannel synthetic aperture radar. *Electronics Communication Engineering Journal, 11*, 29–38.
3. Chapin, E., & Chen, C. W. (2009). Airborne along-track interferometry for GMTI. *IEEE Aerospace and Electronic Systems Magazine, 24*, 13–18.
4. Chapman, R. D., Hawes, C. M., & Nord, M. E. (2010). Target motion ambiguities in single-aperture synthetic aperture radar. *IEEE Transactions on Aerospace and Electronic Systems, 46*, 459–468.
5. Poisson, J. B., Oriot, H. M., & Tupin, F. (2015). Ground moving target trajectory reconstruction in single-channel circular SAR. *IEEE Transactions on Geoscience and Remote Sensing, 53*, 1976–1984.
6. Shen, W., Lin, Y., Zhao, Y., Yu, L., & Hong, W. (2017). Initial result of single channel CSAR GMTI based on background subtraction. In: *2017 IEEE International Geoscience and Remote Sensing Symposium (IGARSS)*, pp. 976–979.
7. Deming, R., Ilin, R., & Best, M. (2013). Phase-wrapping ambiguity in along-track interferometry. In: *2013 Proc. SPIE*, Bellingham, WA (Vol. 8746, p. 87460E).

Ship Detection Using X-Bragg Scattering Model Based on Compact Polarimetric SAR

Chenghui Cao, Xingpeng Mao, Jie Zhang, Junmin Meng, Xi Zhang, and Genwang Liu

Abstract Compact polarimetric SAR is currently drawing more attention owing to its advantage in earth observations. In this paper, based on scattering vector in hybrid mode and X-Bragg scattering model, a new method is presented for evaluating ship detection performance. By using this method, three polarization features, including circular polarization ratio, relative phase, and roundness, were analyzed selectively. Experiments performed using hybrid mode emulated from C-band RADARSAT-2 full polarimetric SAR data validate the feasibility of the method in analyzing the ship detection performance.

Keywords Compact polarimetric SAR · Scattering model · Ship detection · Polarization features

This work was supported by the National Key R&D Program of China (2017YFC1405201), Public Science and Technology Research Funds Projects of Ocean (201505002), and the National Key R&D Program of China (2016YFC1401001).

C. Cao
Harbin Institute of Technology (HIT), Harbin, China

First Institute of Oceanography, Ministry of Natural Resources, Qingdao, China

X. Mao (✉)
Harbin Institute of Technology (HIT), Harbin, China
e-mail: mxp@hit.edu.cn

J. Zhang · J. Meng · X. Zhang · G. Liu
First Institute of Oceanography, Ministry of Natural Resources, Qingdao, China
e-mail: zhangjie@fio.org.cn; mengjm@fio.org.cn; xi.zhang@fio.org.cn; liugenwang@fio.org.cn

E. T. Quinto et al. (eds.), *The Proceedings of the International Conference on Sensing and Imaging, 2018*, Lecture Notes in Electrical Engineering 606,
https://doi.org/10.1007/978-3-030-30825-4_8

1 Introduction

Ship surveillance is of great significance to ocean security and marine management [1]. The synthetic aperture radar (SAR), which can work day and night with high resolution, even under cloudy conditions, has been given great concern for ship detection over the past 30 years.

With the improvement of radar system and ship detection algorithms, the SAR data acquisition mode has been extended from single and dual polarimetric SAR to full polarimetric (FP) SAR [2]. Contrast with single and dual polarimetric SAR, FP SAR can provide more target information and allows complete backscattering characterization of scatterers [3]. However, the narrow swath width of FP SAR can't meet the application demands (*e.g.*, the maximum swath of radarsat-2 is just 50 km), which limits its development.

To solve the problem in FP SAR, compact polarimetric (CP) SAR appears with wide swath coverage and less energy budget [4–6], which has been widely studied, and a lot of promising results have been obtained [5–7]. According to the polarization state, three CP SAR modes have been proposed, including $\pi/4$ [8], dual circular polarization [9], and hybrid polarization (HP) [10]. Among which, the HP mode is simpler, more stable and less sensitive to noise than the other two modes. Furthermore, the HP mode achieves a better performance in self-calibration and engineering [11]. So far, the RISA-1 in India, the ALOS-2 in Japan, even the future Canadian RADARSAT Constellation Mission (RCM) all supports the HP mode, which is most suitable for marine applications.

Extracting effective polarization features from Stokes Vector can essentially reflect the physical difference between ships and sea clutter [6]. The feasibility of feature extraction was verified by researchers in a series of study. For example, Shirvany et al. [12] indicated the effectiveness of the degree of polarization (DoP) in ship detection. Then, this work was further studied by Touzi, who defined an excursion of the DoP to enhance a significant ship-sea contrast [13]. In contrasting with single polarimetric parameter, Yin investigated the capability of m-α and m-χ decompositions for ship detection [14]. Three features extracted from CP SAR were proved to have a good performance in ship detection [3]. What's more, Paes showed a more detailed analysis of the detection capability and sensitivity [15] of δ together with m, μ_C, $|\mu_{xy}|$, and entropy H_ω.

In summary, the existing polarization parameters from polarization decomposition are used directly without considering the influence of sea surface roughness, which weakens the detection capability of the polarization parameters to some extent. In this paper, a new method for analyzing the polarization parameters is presented by introducing the sea surface roughness disturbance [16]. In this case, the polarization parameters can be evaluated in both theory and experiments for ship detection.

Section 2 gives the formula derivation of the presented target scattering model, and the surface roughness of sea is introduced. In Sect. 3, the presented target scattering model is applied to analyze the polarization difference of circular polarization ratio, roundness, and relative phase between ships and sea to prove its availability. Finally, in Sect. 4, the conclusion is given.

2 New Method for Feature Analysis

In hybrid mode, the radar antenna transmits a circular signal and simultaneously receives two orthogonal linear-polarization signals. Consider that radar transmits a right-circular signal, the scattering vector is [17]

$$\vec{k}_l = [E_{\mathrm{RH}} \;\; E_{\mathrm{RV}}]^T = \frac{1}{\sqrt{2}}\left(S_{\mathrm{HH}} - jS_{\mathrm{HV}}, -jS_{\mathrm{VV}} + S_{\mathrm{HV}}\right)^T \quad (1)$$

As we know, the coherency matrix T is a 3×3 matrix [3] as follows:

$$T = \vec{k}_{\mathrm{p}} \vec{k}_{\mathrm{p}}^{\mathrm{H}} = \begin{bmatrix} T_{11} & T_{12} & T_{13} \\ T_{21} & T_{22} & T_{23} \\ T_{31} & T_{32} & T_{33} \end{bmatrix} = \begin{bmatrix} 2A_0 & C - jD & H + jG \\ C + jD & B_0 + B & E + jF \\ H - jG & E - jF & B_0 - B \end{bmatrix} \quad (2)$$

where $\vec{k}_{\mathrm{p}} = \frac{1}{\sqrt{2}}[S_{\mathrm{HH}} + S_{\mathrm{VV}} \;\; S_{\mathrm{HH}} - S_{\mathrm{VV}} \;\; 2S_{\mathrm{HV}}]^T$, $A0$, B, $B0$, C, D, E, F, G, H are Huynen parameters [18].

The matrix T_3 can be expressed by S_{HH}, S_{HV}, S_{VV}, which is extremely complicated. In this case, a new idea is proposed by using the elements of the scattering vector. We do the operation as

$$\begin{cases} E_{\mathrm{RH}} + jE_{\mathrm{RV}} = S_{\mathrm{HH}} + S_{\mathrm{VV}} \\ E_{\mathrm{RH}} - jE_{\mathrm{RV}} = S_{\mathrm{HH}} - S_{\mathrm{VV}} - 2jS_{\mathrm{HV}} \end{cases} \quad (3)$$

Then a new matrix Y is defined consequently as

$$\begin{aligned} Y &= \begin{bmatrix} E_{\mathrm{RH}} + jE_{\mathrm{RV}} \\ E_{\mathrm{RH}} - jE_{\mathrm{RV}} \end{bmatrix} \left[(E_{\mathrm{RH}} + jE_{\mathrm{RV}})^{\mathrm{H}} \;\; (E_{\mathrm{RH}} - jE_{\mathrm{RV}})^{\mathrm{H}} \right] \\ &= \begin{bmatrix} S_{\mathrm{HH}} + S_{\mathrm{VV}} \\ S_{\mathrm{HH}} - S_{\mathrm{VV}} - 2jS_{\mathrm{HV}} \end{bmatrix} \left[(S_{\mathrm{HH}} + S_{\mathrm{VV}})^{\mathrm{H}} \;\; (S_{\mathrm{HH}} - S_{\mathrm{VV}} - 2jS_{\mathrm{HV}})^{\mathrm{H}} \right] \\ &= \begin{bmatrix} (S_{\mathrm{HH}} + S_{\mathrm{VV}})(S_{\mathrm{HH}} + S_{\mathrm{VV}})^* & (S_{\mathrm{HH}} + S_{\mathrm{VV}})\left((S_{\mathrm{HH}} - S_{\mathrm{VV}})^* + 2jS_{\mathrm{HV}}^*\right) \\ (S_{\mathrm{HH}} - S_{\mathrm{VV}} - 2jS_{\mathrm{HV}})(S_{\mathrm{HH}} + S_{\mathrm{VV}})^* & (S_{\mathrm{HH}} - S_{\mathrm{VV}} - 2jS_{\mathrm{HV}})\left((S_{\mathrm{HH}} - S_{\mathrm{VV}})^* + 2jS_{\mathrm{HV}}^*\right) \end{bmatrix} \end{aligned} \quad (4)$$

The matrix Y can be described by the coherency matrix T

$$Y = \begin{bmatrix} T_{11} & T_{12} + jT_{13} \\ T_{12}^* - jT_{13}^* & T_{22} + T_{33} - 2\,\mathrm{Im}\,(T_{23}) \end{bmatrix} \quad (5)$$

By using Huynen parameters [18] the matrix Y is

$$Y = \begin{bmatrix} 2A_0 & (C-G) + j(H-D) \\ (C-G) - j(H-D) & 2(B_0 - F) \end{bmatrix} \tag{6}$$

In Wang et al. [18], the stokes vector of the scattered wave in hybrid mode is written as

$$g = \begin{bmatrix} g_0 \\ g_1 \\ g_2 \\ g_3 \end{bmatrix} = \begin{bmatrix} A_0 + B_0 - F \\ C - G \\ H - D \\ -A_0 + B_0 - F \end{bmatrix} \tag{7}$$

Hence, Y can be derived from Eqs. (5), (6), and (7) as

$$\begin{aligned} Y &= \begin{bmatrix} T_{11} & T_{12} + jT_{13} \\ T_{12}^* - jT_{13}^* & T_{22} + T_{33} - 2\,\mathrm{Im}(T_{23}) \end{bmatrix} \\ &= \begin{bmatrix} g_0 - g_3 & g_1 + jg_2 \\ g_1 - jg_2 & g_0 + g_3 \end{bmatrix} \end{aligned} \tag{8}$$

As a result, the stokes vector can be derived as

$$\begin{cases} g_0 = \frac{T_{11}+T_{22}+T_{33}}{2} - \mathrm{Im}(T_{23}) \\ g_1 = \mathrm{Re}(T_{12}) - \mathrm{Im}(T_{13}) \\ g_2 = \mathrm{Im}(T_{12}) + \mathrm{Re}(T_{13}) \\ g_3 = \frac{-T_{11}+T_{22}+T_{33}}{2} - \mathrm{Im}(T_{23}) \end{cases} \tag{9}$$

Based on the theory mentioned above, the coherency matrix T and stokes vector are used to represent the constructed matrix Y. For better description of the matrix Y, X-Bragg scattering model is introduced below.

X-Bragg scattering model is first introduced by Hajnsek et al. [16] in order to solve the case of nonzero cross-polarized backscattering and depolarization. By assuming a roughness disturbance induced random surface slope β, the X-Bragg scattering is modeled as a reflection depolarizer [16] as shown in Eq. (10).

$$T_{\text{X-Bragg}} = \begin{bmatrix} T_{11} & T_{12} & T_{13} \\ T_{21} & T_{22} & T_{23} \\ T_{31} & T_{32} & T_{33} \end{bmatrix} = \begin{bmatrix} C_1 & C_2 \sin c(2\beta_1) & 0 \\ C_2 \sin c(2\beta_1) & C_3(1 + \sin c(4\beta_1)) & 0 \\ 0 & 0 & C_3(1 - \sin c(4\beta_1)) \end{bmatrix} \tag{10}$$

where

$$\begin{cases} C_1 = |R_S + R_P|^2 \\ C_2 = (R_S + R_P)\left(R_S^* - R_P^*\right) \\ C_3 = \frac{1}{2}|R_S - R_P|^2 \end{cases}$$

β is assumed to follow a uniform distribution in the interval $[-\beta 1, \beta 1]$, and $\beta 1 < \pi/2$ is the distribution width.

Combined with Eqs. (9) and (10), the stokes vector can be described by X-Bragg scattering matrix as

$$\begin{cases} g_0 = \frac{T_{11}+T_{22}+T_{33}}{2} - \mathrm{Im}\,(T_{23}) = \frac{C_1+2C_3}{2} \\ g_1 = \mathrm{Re}\,(T_{12}) - \mathrm{Im}\,(T_{13}) = \mathrm{Re}\,(C_2 \sin c\,(2\beta_1)) \\ g_2 = \mathrm{Im}\,(T_{12}) + \mathrm{Re}\,(T_{13}) = \mathrm{Im}\,(C_2 \sin c\,(2\beta_1)) \\ g_3 = \frac{-T_{11}+T_{22}+T_{33}}{2} - \mathrm{Im}\,(T_{23}) = \frac{-C_1+2C_3}{2} \end{cases} \tag{11}$$

With the analysis of scattering difference of ships and sea, Eq. (11) can evaluate the performance of the polarization features in ship detection.

3 Ship Detection Performance Analysis

According to the target scattering model of ships in Sect. 2, the polarization parameters, such as, circular polarization ratio, roundness, and relative phase, which have good performance in ship detection [19], are analyzed for example to extract new parameter. For lack of compact polarimetric SAR data, full polarimetric RADARSAT-2 SAR data are chosen to reconstruct the compact polarimetric SAR data transmitted by right-circular polarization and received by horizontal and vertical polarization. Note that the formulas are

$$\mathrm{CPR} = \frac{g_0 - g_3}{g_0 + g_3}, \quad \sin 2\chi = -\frac{g_3}{\sqrt{g_1^2 + g_2^2 + g_3^2}}, \quad \delta = \tan^{-1}\left(\frac{g_3}{g_2}\right) \tag{12}$$

Circular Polarization Ratio

The circular polarization ratio (CPR) is derived from Eqs. (11) and (12) as

$$\mathrm{CPR} = \frac{C_1}{2C_3} = \frac{|R_S + R_P|^2}{|R_S - R_P|^2} = \frac{|S_{HH} + S_{VV}|^2}{|S_{HH} - S_{VV}|^2} \tag{13}$$

In Eq. (13), the value of the CPR is positive, which is only related to the dielectric constant and the incidence angle, and is independent of the rotation angle. Therefore, this parameter is stable and can be used in ship detection.

Figure 1 is a part of SAR image, which shows the value of CPR about ships and sea. We can see the value of the ship is basically between 0 and 1, while the value of the sea surface is greater than 1. Hence, the ships and sea surface can be separated

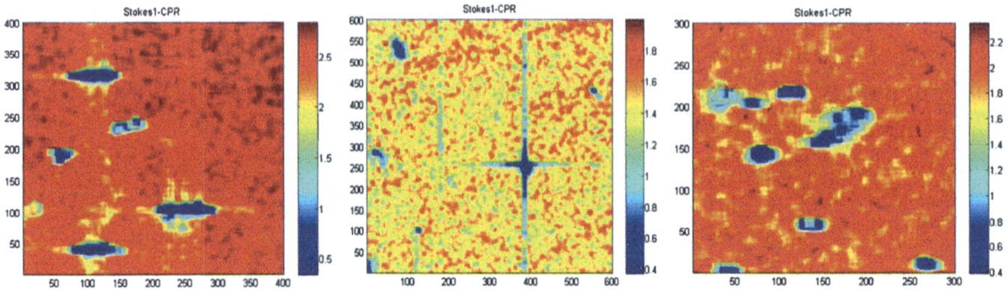

Fig. 1 Value of circular polarization ratio about the ships and sea surface

by the threshold 1. However, the value of CPR in the figure ranges from 0.4 to 2.5, which means the span is too small to select the threshold properly.

Roundness

Roundness can be derived from Eqs. (11) and (12) as

$$\sin 2\chi = \frac{C_1 - 2C_3}{\sqrt{4(C_2 \sin c\,(2\beta_1))^2 + (-C_1 + 2C_3)^2}} \tag{14}$$

In Eq. (14), the value of χ and sin2χ is consistent whether it is positive or negative. On the right side of the Eq. (14), the denominator is positive, so the positive and negative properties of χ depends on the numerator. Assuming that the phase angle of $S_{\mathrm{HH}}S_{\mathrm{VV}}^{*}$ is $\theta_{\mathrm{HH-VV}}$, the numerator is derived as

$$C_1 - 2C_3 = 2\,\mathrm{Re}\left(S_{\mathrm{HH}}S_{\mathrm{VV}}^{*}\right) = 2\cos\left(\theta_{\mathrm{HH\text{-}VV}}\right) \tag{15}$$

According to Eq. (15), the denominator on the right side is positive, as a result, the value of $C_1 - 2C_2$ is decided by the phase angle $\theta_{\mathrm{HH-VV}}$. We know that the initial value of the phase angle is $\pi/2$, and with each additional scattering, the phase angle increases π. Therefore, the phase angles are $3\pi/2$ and $5\pi/2$ for single scattering and even scattering respectively. Consequently, for single scattering, the value of cos($\theta_{\mathrm{HH-VV}}$) is positive, while for even scattering, the value is negative.

The scattering of sea surface is mostly surface scattering, and the ship is mostly even scattering, so the value of sea surface should be positive and the ship should be negative. As is shown in Fig. 2, the ships and sea surface can be separated by the threshold 0. Notice that the value of roundness is related to the rotation angle, which is not stable in high sea conditions.

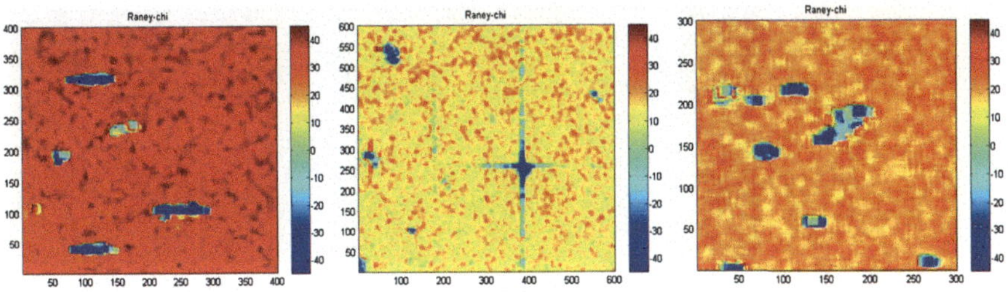

Fig. 2 Value of roundness about the ships and sea surface

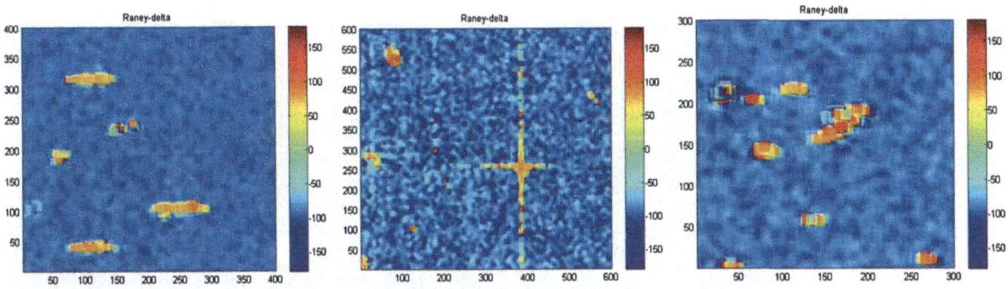

Fig. 3 Value of relative phase about the ships and sea surface

Relative Phase

Relative phase is derived from Eqs. (11) and (12) as

$$\delta = -\frac{1}{\sin c\,(2\beta_1)} \cot\left(\theta_{\text{HH-VV}}\right) \tag{16}$$

In Eq. (16), for single scattering, the phase angle is $3\pi/2$, so the value of $\cot(\theta_{\text{HH - VV}})$ is positive, which means the value of δ is negative. While for even scattering, the value of δ is positive. Due to different scattering characteristics, the ships in SAR image is mainly even scattering, while the sea surface is mainly surface scattering of ships and sea surface. Therefore, the value of the ships should be positive and the sea surface should be negative. An example is shown in Fig. 3, and the threshold 0 can be used to separate the ships and sea surface.

Note that the value of δ is related to the angle β in Eq. (16), the surface roughness increases with the increasing sea conditions. Therefore, the value of δ is unstable influenced by β, which is difficult to distinguish ships and the sea surface, especially in high sea conditions.

Figures 4 and 5 are experiments' results: (a) is the amplitude of RV polarization, and (b), (c), and (d) are results of threshold segmentation (the threshold are 1, 0 and 0). The areas with white are ships while dark and gray are sea surface. It indicates that the three features are all good discriminators for observing ships from sea surface. According to the scattering intensity, ships in Fig. 5 are bigger than

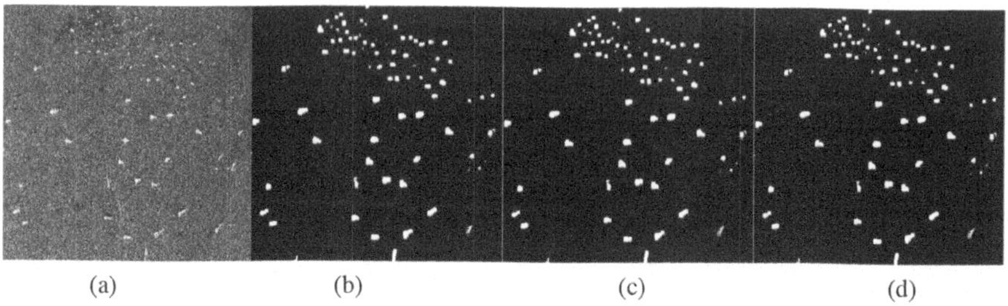

Fig. 4 Results of threshold segmentation. (**a**) RV image (**b**) CPR (**c**) Roundness (**d**) Relative phase

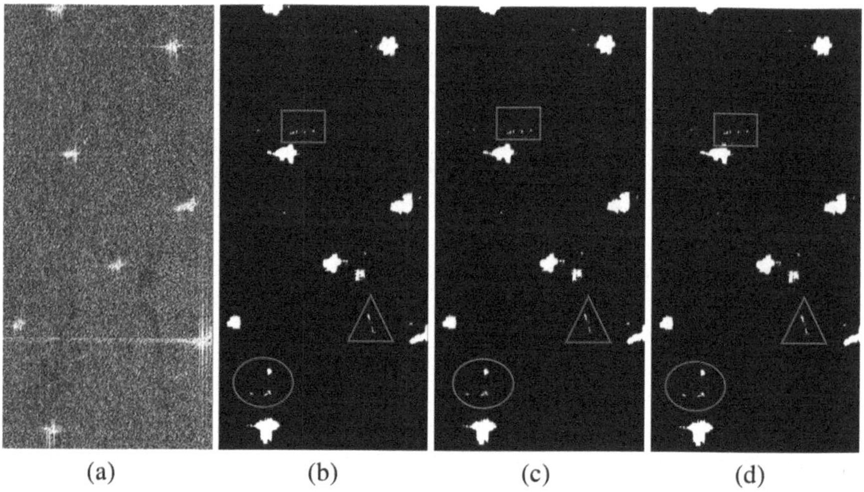

Fig. 5 Results of threshold segmentation. (**a**) RV image (**b**) CPR (**c**) Roundness (**d**) Relative phase

ships in Fig. 4. Combined with AIS, false alarms circled by red box are apparently owing to the high sea condition in Fig. 5. All in all, roundness and relative phase are all unstable in high sea conditions, while for CPR, the proper threshold is hard to choose in ship detection. The total number of detected ships is 166, and the false alarm rate of CPR, roundness, and relative phase are 0.06, 0.078, and 0.083, respectively.

4 Conclusion

In this paper, a new method of analyzing polarization parameter is derived by introducing the surface roughness of the sea. Three compact polarimetric parameters in hybrid mode are analyzed. The circular polarization ratio, roundness, and relative phase can all be used to separate ships from sea surface. The circular polarization ratio is stable, but the span value is too small to select proper threshold for ship detection. Roundness and relative phase are not stable owing to their relation with rotation angle β, which increases the difficulty of ship detection, especially in high sea conditions. The false alarm rate of CPR, roundness, and relative phase are 0.06, 0.078, and 0.083, respectively by detecting 166 ships.

References

1. Ouchi, K. (2016). Current status on vessel detection and classification by synthetic aperture radar for maritime security and safety (pp. 5–12). In *Proceedings of Symposium on Remote Sensing of Environmental Sciences*, Gamagori, Japan, 2016.
2. Gao, G., Gao, S., He, J., et al. (2018). Adaptive ship detection in hybrid-polarimetric SAR images based on the power-entropy decomposition. *IEEE Transactions on Geoscience & Remote Sensing, 99*, 1–14.
3. Yin, J., Yang, J., Zhou, Z. S., et al. (2015). The extended bragg scattering model-based method for ship and oil-spill observation using compact polarimetric SAR. *IEEE Journal of Selected Topics in Applied Earth Observations & Remote Sensing, 8*(8), 3760–3772.
4. Souyris, J. C., & Sandra, M. (2002). Polarimetry based on one transmitting and two receiving polarizations: the pi/4 mode (pp. 629–631). In *2002 IEEE International Geoscience and Remote Sensing Symposium, IGARSS'02*, 2002.
5. Charbonneau, F. J., Brisco, B., Raney, R. K., et al. (2010). Compact polarimetry overview and applications assessment. *Canadian Journal of Remote Sensing, 36*(Suppl 2), S298–S315.
6. Raney, R. K. (2006). Dual-polarized SAR and stokes parameters. *IEEE Geoscience and Remote Sensing Letters, 3*(3), 317–319.
7. Ouchi, K. (2013). Recent trend and advance of synthetic aperture radar with selected topics. *Remote Sensing, 5*(2), 716–807.
8. Souyris, J. C., Imbo, P., Fjortoft, R., et al. (2005). Compact polarimetry based on symmetry properties of geophysical media: The/spl pi//4 mode. *IEEE Transactions on Geoscience and Remote Sensing, 43*(3), 634–646.
9. Stacy, N., & Preiss, M. (2006). Compact polarimetric analysis of X-band SAR data (p. 6). In *Proceeding of EUSAR*, 2006.
10. Raney, R. K. (2007). Hybrid-polarity SAR architecture. *IEEE Transactions on Geoscience and Remote Sensing, 45*(11), 3397–3404.
11. Gao, G., Gao, S., He, J., et al. (2018). Ship detection using compact polarimetric SAR Based on the notch filter. *IEEE Transactions on Geoscience & Remote Sensing, 99*, 1–14.
12. Shirvany, R., Chabert, M., & Tourneret, J. Y. (2012). Ship and oil-spill detection using the degree of polarization in linear and hybrid/compact dual-pol SAR. *IEEE Journal of Selected Topics in Applied Earth Observations and Remote Sensing, 5*(3), 885–892.
13. Touzi, R., & Vachon, P. W. (2015). RCM polarimetric SAR for enhanced ship detection and classification. *Canadian Journal of Remote Sensing, 41*(5), 473–484.
14. Yin, J., & Yang, J. (2014). Ship detection by using the M-Chi and M-Delta decompositions (pp. 2738–2741). In *Geoscience and Remote Sensing Symposium. IEEE*, 2014.
15. Paes, R. L., Nunziata, F., & Migliaccio, M. (2016). On the capability of hybrid-polarity features to observe metallic targets at sea. *IEEE Journal of Oceanic Engineering, 41*(2), 346–361.
16. Hajnsek, I., Pottier, E., & Cloude, S. R. (2003). Inversion of surface parameters from polarimetric SAR[J]. *IEEE Transactions on Geoscience and Remote Sensing, 41*(4), 727–744.

17. Cloude, S. R., Goodenough, D. G., & Chen, H. (2012). Compact decomposition theory. *IEEE Geoscience and Remote Sensing Letters, 9*(1), 28–32.
18. Wang, H., Zhou, Z., Turnbull, J., et al. (2015). Three-component decomposition based on stokes vector for compact polarimetric SAR. *Sensors, 15*(9), 24087–24108.
19. Cao, C.-H., Zhang, J., Zhang, X., et al. (2017). The analysis of ship target detection performance with C band compact polarimetric SAR. *Periodical of Ocean University of China, 47*(2), 85–93.

Analysis of Sea Clutter Distribution and Evaluation of Ship Detection Performance for Sentinel-1 SAR Data

Yongxu Li, Xudong Lai, Jie Zhang, Junmin Meng, Genwang Liu, and Xi Zhang

Abstract This paper statistically analyzed the sea clutter distribution and ship detection performance for Sentinel-1 synthetic aperture radar image. First, the goodness-of-fit of five commonly used distribution models were evaluated to find out the most suitable model and the Kullback–Leibler Distance was adopted to judge the fitting degree. Then construct a constant false alarm rate detector for ship detection. To measure the robustness of the detector, the figure of merit, the probability of detection, and the false alarm rate were calculated to evaluate detector performance.

Keywords Ship detection · Synthetic aperture radar · Constant false alarm rate · Statistical distribution

1 Introduction

The monitoring of the ship target is one of the most important research fields. The information such as location and type of ship is widely used in maritime surveillance, maritime detection, traffic safety, fisheries control, and so on. For synthetic aperture radar (SAR), the basic method of ship detection takes advantage of SAR images' feature that the backscattering signal from the ship is much stronger

Y. Li (✉)
The First Institute of Oceanography, Ministry of Natural Resources, Qingdao, China

School of Remote Sensing and Information Engineering, Wuhan University, Wuhan, China
e-mail: Liyongxu@whu.edu.cn

X. Lai
School of Remote Sensing and Information Engineering, Wuhan University, Wuhan, China
e-mail: laixudong@whu.edu.cn

J. Zhang · J. Meng · G. Liu · X. Zhang
The First Institute of Oceanography, Ministry of Natural Resources, Qingdao, China
e-mail: zhangjie@fio.org.cn; mengjm@fio.org.cn; liugenwang@fio.org.cn; xi.zhang@fio.org.cn

E. T. Quinto et al. (eds.), *The Proceedings of the International Conference on Sensing and Imaging, 2018*, Lecture Notes in Electrical Engineering 606,
https://doi.org/10.1007/978-3-030-30825-4_9

than the sea clutter background in most cases, and the detection was realized by searching pixels whose amplitudes are greater than a given threshold. The constant false alarm rate (CFAR) detector was widely used due to the variable threshold, which was determined by accurately describing the real-time dynamic sea clutter around the target. An appropriate model is essential to the detector [1–3].

However, properties like frequency, polarization, resolution, grazing angle, and sea state have influence on SAR imaging, therefore, the sea clutter may fit different models and the suitable model of the given dataset needs to be judged [4–8]. In high-resolution SAR image, the K distribution becomes popular due to the compound formulation, which was introduced by Ward that enables both the small-scale and large-scale components of the sea clutter to be characterized [9, 10]. The Weibull (WBL) distribution was used to model amplitude earlier, and the results show that it can fit most SAR images effectively. The lognormal (LGN) distribution can achieve better goodness-of-fit even under the heterogeneous situation in high-resolution SAR images [11]. The computational complexity of G^0 distribution was reduced by eliminating an iterative computing step; meanwhile, it has been demonstrated with excellent performance in a heterogeneous sea surface environment [12]. The generalized gamma distribution (GΓD) was used for modeling many scenes of high-resolution SAR images and shows a better performance in most cases [13].

Therefore, the comparative analyses among those five commonly used distributions were carried out. The Kullback–Leibler (K–L) Distance was used to judge the goodness-of-fit. After the best fitting was found, the CFAR detector was constructed to carry out ship detection experiments. To evaluate CFAR detector performance, the ground truth (GT), which were determined via Automatic Identification System (AIS) data combined with manual interpretation, was used to validate the result, and the figure of merit (FOM), probability of detection (PoD), and false alarm rate (FAR) were calculated.

The remainder of this paper is organized as follows. Section 2 briefly describes the Sentinel-1 dataset. The experiment methodologies are presented in Sect. 3, while Sect. 4 shows the results. Finally, the conclusions are given in Sect. 5.

2 Sentinel-1 IW Level-1 GRD Data

Following the "open and free" data access policy, seven dual polarization images of Sentinel-1 IW Level-1 Ground Range Detected High resolution (GRDH) produce from October 2014 to January 2016 have been obtained in the three areas of the Strait of Malacca, shown in Fig. 1. These images were acquired using vertical transmit, vertical receive (VV) polarization and vertical transmit, horizontal receive (VH) polarization in high resolution 20 m × 22 m.

In order to avoid the influence of shoreline or island on the experimental results, 30 pairs of sub-images were extracted from the original data by artificial cutting. In addition to the goodness-of-fit analysis, whether the detector has the capability to provide reliable results is also the scope of the assessment. We verified 304 ship

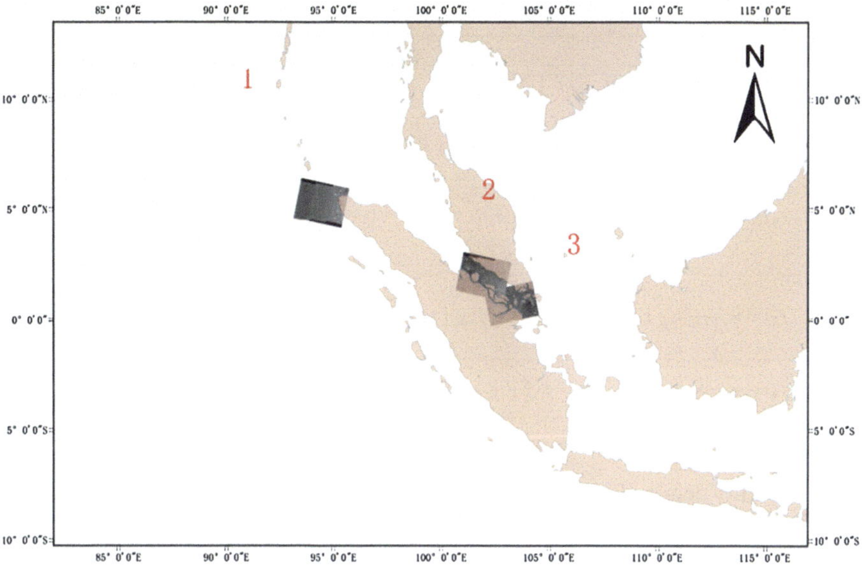

Fig. 1 Geographical location

Table 1 Additional information

No.	Acquire time	Location	GT	Sub-images
1	20150815_232741	1	9	2
2	20151026_232742	1	5	1
3	20141222_225535	2	74	6
4	20150316_225534	2	72	6
5	20150527_225538	2	69	6
6	20151018_225543	2	58	6
7	20160116_112449	3	17	3

targets by AIS data and manual supervision to acquire GT as the reference. To acquire GT, 304 ship targets were verified by AIS data and manual supervision. Additional information about the dataset is shown in Table 1.

3 Experiment Methodologies

Distribution Model

In this section, these five distribution models: LGN, WBL, *K*-root, GΓD, G^0 were used to fit the sea clutter. After the distribution model was determined, the Method of log-cumulants (MoLC) was used to solve the parameters of the distribution model by counting the pixels in the background box around the target to obtain

Table 2 Amplitude PDFs and MoLC equations of the models

Model	PDF	MoLC equations
WBL	$f_A(x) = \frac{\gamma}{\sigma^{\gamma}}(x)^{\gamma-1}\exp\left[-\left(\frac{x}{\sigma}\right)^{\gamma}\right], x, \gamma, \sigma > 0$	$k_1 = \log(\sigma) + \Phi_0(1)\gamma^{-1}$ $k_2 = \Phi_0(1, 1)\gamma^{-2}$
LGN	$f_A(x) = \frac{1}{\sqrt{2\pi}\sigma x}\exp\left[-\frac{(\ln x - \mathrm{m})^2}{2\sigma^2}\right], \sigma > 0, m \in \mathrm{R}$	$k_1 = m$ $k_2 = \sigma^2$
G^0	$f_A(x) = \frac{2L^L(L-\alpha)}{\Gamma^{\alpha}\Gamma(L)\Gamma(-\alpha)}\frac{x^{2L-1}}{\left(\gamma+Lx^2\right)^{L-\alpha}}, x, L, \gamma > 0, \alpha < 0$	$2k_1 = \log\gamma/L + \Phi_0(L) - \Phi_0(-\alpha)$ $4k_2 = \Phi_0(1, L) + \Phi_0(1, -\alpha)$ $8k_3 = \Phi_0(2, L) - \Phi_0(2, -\alpha)$
K-root	$f_A(x) = \frac{4}{\Gamma(L)\Gamma(\gamma)}\left(\frac{L\gamma}{\mu}\right)^{\frac{L+\gamma}{2}}x^{L+\gamma-1}K_{\lambda-L}\left(2x\sqrt{\frac{L\gamma}{\mu}}\right), x > 0$	$2k_1 = \log\mu/L\gamma + \Phi_0(L) + \Phi_0(\gamma)$ $4k_2 = \Phi_0(1, L) + \Phi_0(1, \gamma)$ $8k_3 = \Phi_0(2, L) + \Phi_0(2, \gamma)$
GΓD	$f_A(x) = \frac{\lvert\gamma\rvert\kappa^{\kappa}}{\sigma\Gamma(\kappa)}\left(\frac{x}{\sigma}\right)^{\kappa\gamma-1}\exp\left[-\kappa\left(\frac{x}{\sigma}\right)^{\gamma}\right], x \in \mathrm{R}^{+}$	$k_1 = \log(\sigma) + \left(\Phi_0(\kappa) - \log(\kappa)\right)\gamma^{-1}$ $k_i = \Phi_0(\mathrm{i}-1, \kappa)\gamma^{-i}, i = 2, 3\ldots$

the probability density function (PDF) of the distribution models. The PDFs and the corresponding equations for the models are shown in Table 2 [14].

CFAR Detector

The CFAR detector is widely used in ship detection by comparing pixels with an adaptive threshold **T** to maintain a constant false alarm rate. **T** can be obtained by solving Eq. (1), according to PDF $f(x)$ and the desired probability of false alarm (p_{fa}). The former was obtained by modeling the distribution of sea clutter from the background cell, and the latter was based on experience [15].

$$p_{\mathrm{fa}} = 1 - \int_0^T f(x)dx = \int_T^{\infty} f(x)dx \tag{1}$$

Evaluation Methodology

To evaluate the fitting degree of different models, the K–L Distance was adopted to access the modeling accuracy, which is a measure of how one probability distribution is different from another reference, reference probability distribution PDF, and histogram of sea clutter in this case. The lower the K–L Distance, the higher the fitting accuracy.

$$I(g, f) = \int g(x)\ln\frac{g(x)}{f(x)}dx \tag{2}$$

where $f(x)$ is the PDF and $g(x)$ is the histogram of sea clutter.

The detector's performance was quantitatively evaluated by comparing results with GT, and the following metrics were estimated:

FOM:

$$\mathrm{FOM} = N_{\mathrm{tt}} / \left(N_{\mathrm{gt}} + N_{\mathrm{fa}} \right) \quad (3)$$

PoD:

$$\mathrm{PoD} = \left(N_{\mathrm{tt}} / N_{\mathrm{gt}} \right) \quad (4)$$

FAR:

$$\mathrm{FAR} = \left(N_{\mathrm{fa}} / N_{\mathrm{gt}} \right) \quad (5)$$

where N_{tt} is the total number of detection targets matching GT, N_{gt} is the total number of GT, and N_{fa} is the total number of false alarms.

4 Results

Analysis of Goodness-of-Fit

The results of goodness-of-fit test were illustrated in Fig. 2. The vertical coordinate recorded the K–L Distance, the transverse coordinate indicates sample number that was sorted by incident angle; results under VV are shown in Fig. 2a and under VH are shown in Fig. 2b. As shown, there was no obvious correlation between incident angles with fitting accuracy. However, there are significant differences among polarizations. That is, for the same distribution model, the results under VH were significantly lower and were more uniform than those under VV. Furthermore, analyzing the differences between models, the other four distributions except WBL can hardly be discerned from the diagram.

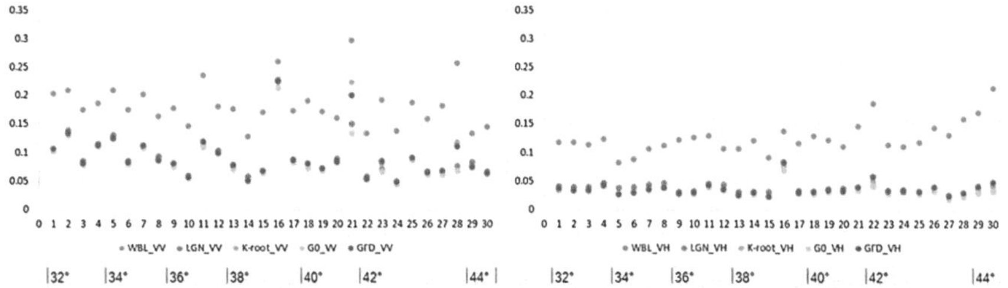

Fig. 2 K–L Distance of distributions under VV (left plot), VH (right plot)

To quantify the experimental results and find out the most accurate and robust distribution, the mean and the variance value of those results were counted and collected in Table 3. Take LGN as an example, the mean and the variance value of VH were much smaller than those of VV. The former exhibited values of 0.037 and 0.009, while for the latter, they were 0.091 and 0.035, respectively. That means the fitting performance under VH is better than that of VV.

Moreover, mark the best performance of each indicator in bold. As can be seen, the bold indicators are clustered in the third column of the table, which were the results for the G^0. That means G^0 has the best fitting with an overall average of 0.058 and an overall variance of 0.036. The results of LGN, K-root, and GΓD are basically identical, which are clearly better than the WBL. Therefore, G^0 was used to construct a CFAR detector for ship detection experiments and the detector performance was evaluated below.

Detection Performance

The parameter that controls the performance of CFAR detectors is the p_{fa}. By recording the results of the detector under different p_{fa}, the curves of the FOM, the PoD, and the FAR changing with p_{fa} were drawn, with transverse coordinate recording the values of p_{fa} (10^{-x}), and vertical coordinate recording the statistical results of detector's FOM, PoD, FAR, respectively. The performance of the detector was directly shown in Fig. 3.

Table 3 The mean and the variance of K–L distance

	LGN	G^0	K-root	GΓD	WBL
VV_mean	0.091	**0.085**	0.092	0.093	0.184
VH_mean	0.037	**0.032**	0.033	0.035	0.125
ALL_mean	0.064	**0.058**	0.063	0.065	0.155
VV_variance	0.035	**0.033**	0.042	0.039	0.038
VH_variance	0.009	**0.009**	0.010	0.011	0.027
ALL_variance	0.037	**0.036**	0.042	0.041	0.044

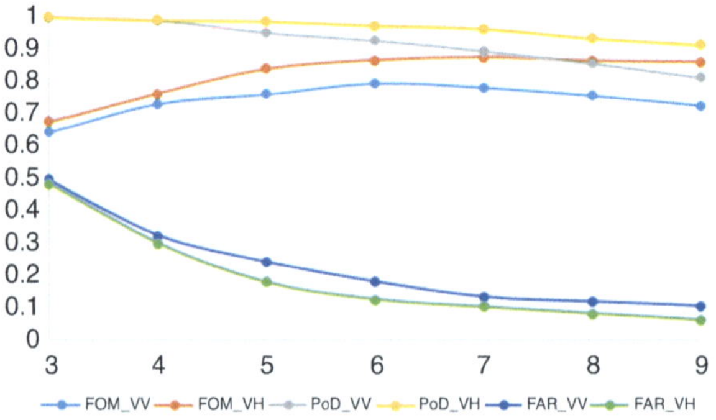

Fig. 3 The FOM, POD, FAR curves of the detector

As can be seen, the results under VH were different from VV, having higher FOM, PoD, and lower FAR, which means the detection performance of VH was better than VV.

In general, with the decrease of p_{fa}, the decreasing trend of PoD under VH was gentle, from 100 to 91.45%, down 8.55%. In contrast, FAR decreased significantly, from 48.36 to 6.25%, down 42.11%. FOM increased from 67.41 to 86.07%, rising 18.66%. However, there was no obvious change after p_{fa} was 10^{-7}. In the case of VV, the decreasing trend of PoD was greater than that of VH, from 99.79 to 81.06%, down 18.72%. Meanwhile, FAR decreased more obviously, from 55.96 to 11.91%, down 44.04%, FOM increased from 63.98% to the highest 79.34% when p_{fa} was 10^{-6}, rising 15.36%, then there was a slight decline from the peak value to 72.43%, down 6.91%.

In conclusion, the CFAR detector based on G^0 distribution performed notably well. The best performance of VH was when the p_{fa} value was 10^{-7}, and so of VV was when the p_{fa} value was 10^{-6}, which has reference significance for the reasonable value of p_{fa}.

5 Conclusions

In this paper, five commonly used distribution models were used to fit the Sentinel-1 SAR image, and the K–L Distance was calculated to quantitatively judge the matching degree. G^0 distribution was shown most fitting to the data, followed by LGN, K-root, GΓD, and WBL. Therefore, a CFAR detector was constructed with G^0 to carry out ship detection experiments, and the results showed that VH was more suitable for ship detection than VV. For VH, the recommended p_{fa} was 10^{-7}. For VV, the recommended p_{fa} was 10^{-6}.

Acknowledgement The authors are very grateful to the European Space Agency for providing the experimental dataset. This work was supported by the Public Science and Technology Research Funds Projects of Ocean under grant 201505002.

References

1. Pelich, R., Longepe, N., Mercier, G., Hajduch, G., & Garello, R. (2015). *Performance evaluation of Sentinel-1 data in SAR ship detection*. IGARSS 2015—2015 IEEE International Geoscience and Remote Sensing Symposium. IEEE.
2. Daum, F. (2008). Radar handbook, 3rd Ed. (M. I. Skolnik, Ed.) [book review]. *IEEE Aerospace & Electronic Systems Magazine, 23*(5), 41–41.
3. Eldarymli, K., Mcguire, P., Power, D., & Moloney, C. R. (2013). Target detection in synthetic aperture radar imagery: A state-of-the-art survey. *Journal of Applied Remote Sensing, 7*(7), 071598–071598.
4. Fiche, A., Angelliaume, S., Rosenberg, L., & Khenchaf, A. (2018). *Statistical analysis of low grazing angle high resolution X-band SAR sea clutter*. Radar Conference. IEEE.

5. Fiche, A., Angelliaume, S., Rosenberg, L., & Khenchaf, A. (2015). Analysis of x-band SAR sea-clutter distributions at different grazing angles. *IEEE Transactions on Geoscience and Remote Sensing, 53*(8), 4650–4660.
6. Xin, Z., Liao, G., Yang, Z., Zhang, Y., & Dang, H. (2017). Analysis of distribution using graphical goodness of fit for airborne SAR sea-clutter data. *IEEE Transactions on Geoscience and Remote Sensing, 55*, 5719–5728.
7. Guillaume, H. (2016). AIS-based evaluation of target detectors and SAR sensors characteristics for maritime surveillance. *IEEE Journal of Selected Topics in Applied Earth Observations & Remote Sensing, 8*(8), 3892–3901.
8. Gao, G. (2010). Statistical modeling of SAR images: A survey. *Sensors (14248220), 10*(1), 775–795.
9. Jakeman, E., & Pusey, P. (2003). A model for non-Rayleigh sea echo. *IEEE Transactions on Antennas and Propagation, 24*(6), 806–814.
10. Ward, K., Tough, R., Watts, S., Ward, K., Tough, R., & Watts, S. (2013). Sea clutter: Scattering, the k-distribution and radar performance. *Waves in Random & Complex Media, 17*(2), 233–234.
11. Goldstein, G. B. (1973). False-alarm regulation in log-normal and Weibull clutter. *IEEE Transactions on Aerospace and Electronic Systems, AES-9*(1), 84–92.
12. Jung, C. H., Yang, H. J., & Kwag, Y. K. (2009). *Local cell-averaging fast CFAR for multi-target detection in high-resolution SAR images*. Asian-pacific Conference on Synthetic Aperture Radar.
13. Qin, X., Zhou, S., Zou, H., et al. (2013). A CFAR detection algorithm for generalized gamma distributed background; in high-resolution SAR images. *IEEE Geoscience & Remote Sensing Letters, 10*(4), 806–810.
14. Cui, S., Schwarz, G., & Datcu, M. (2014). A comparative study of statistical models for multilook SAR images. *IEEE Geoscience and Remote Sensing Letters, 11*(10), 1752–1756.
15. Ji, Y., Zhang, J., Meng, J., & Zhang, X. (2010). A new CFAR ship target detection method in SAR imagery. *Acta Oceanologica Sinica, 29*(1), 12–16.

Reflectance and Surface Normals from a Single Hyperspectral Image

Wenhao Xiang, Xudong Jin, and Yanfeng Gu

Abstract Hyperspectral imaging, which collects rich spectral and spatial information, is a powerful Earth vision method and has many applications. As the data structure is highly complex, the key problem of hyperspectral image processing is in extracting the useful information we want. Traditional feature extraction methods are designed to this end; however, they undergo severe limitations. Most of them are designed mathematically instead of physically and ignore the fact that the changes in physical imaging conditions have a significant influence on the spectra intensity observed. In this chapter, we try to analyze the information contained in hyperspectral images (HSIs) from the perspective of the hyperspectral imaging principle, and propose a novel method of extracting reflectance and surface normals from HSIs.

Keywords Hyperspectral · Reflectance · Surface Normals

1 Introduction

The intrinsic image decomposition (IID) problem is a very important and challenging subject in computer vision and in signal processing. As we can infer from the name, IID is aimed at decomposing an image, which is opposite to the process of acquiring an image. When acquiring an image, we use sensors to scan the surface of the Earth, and obtain a grid-structured 2D array of digital numbers to represent the spectral intensity. This process is known as the "imaging process," which is the result of a complex physical interaction of environment elements, e.g., illumination, surface geometry, surface reflectance, and view position. On the contrary, the IID

W. Xiang
System Engineering Research Institute, China State Shipbuilding Corporation, Beijing, China

X. Jin · Y. Gu (✉)
Harbin Institute of Technology, Harbin, China
e-mail: guyf@hit.edu.cn; guyf@hit.edu.com

E. T. Quinto et al. (eds.), *The Proceedings of the International Conference on Sensing and Imaging, 2018*, Lecture Notes in Electrical Engineering 606,
https://doi.org/10.1007/978-3-030-30825-4_10

process means the "reserve imaging process," i.e., given an hyperspectral image (HSI), how do we recover the corresponding physical scene properties that render the very image. According to the inputs of the IID algorithm, existing IID methods described in the literature can be categorized into four classes: (1) IID methods for a single-segmented object. (2) IID methods for a single image [1]. (3) IID methods for sequences of images [2]. (4) IID methods for RGB-D images [3]. Moreover, in terms of their outputs, IID methods can be categorized as: (1) Those generating the reflectance component and the shading component. (2) Those generating reflectance, shape, and illumination [4]. The IID problem of recovering scene reflectance and surface normals from a single real-world HSI is considered in this chapter.

Different from the traditional IID methods, which only extract reflectance and shading components from HSIs, we considered a more general case that further extracted illumination and surface normals from a shading component. As we know, the shading component describes the phenomenon of the complex interaction between incident light and surface normals. In traditional IID tasks, people usually assume that only the intrinsic component, which describes the intrinsic properties of the material itself, i.e., the reflectance, contains useful information, whereas the shading component, which contains nothing but noise, disturbance and redundancy, should be removed. However, this observation only tells part of the truth. It is true that the shading component contains noises. But it also provides other useful information such as the imaging environment condition and surface topographical structures. With this in mind, we propose a novel framework that treats the shading component as an organic constituent part.

Yet, it is not trivial to recover surface normals from the shading component. First, the shading component only contains low-frequency geometry cues, which are not sufficient because of the problem of predicting high-frequency geometry details. Second, the low spatial resolution result from the small ground sample distance further increases the difficulty. Third, different objects have abrupt height changes, such as that of a building roof to the ground, which is hard to judge where and how intense such changes occur. Taking all these conditions into account, our method makes full use of the spectral information of hyperspectral images. To be more precise, each wavelength band of an HSI consists of highly correlate information on the one hand, and some band-specific information on the other hand, which can provide additional priors to constrain construction.

One possible application of our method is to increase the performance of HSI classifiers. As we all know, it is important to design appropriate features to obtain satisfactory classification results for HSIs. Traditionally, these designed features only describe the implicit data properties, which are highly dependent upon the data parameters, for HSIs with different spatial and spectral images, or HSIs acquired by different sensors or illumination conditions, new features would be desired and need to redesigned. This would cause much inconvenience and more robust features are desired. Different from those implicit features, our method can obtain varying explicit intrinsic properties, i.e., the reflectance, which describes the material properties, and the surface normals, which describe the surface topographical structures. These physical quantities are invariant regardless of imaging conditions or sensor parameters; they are also independent of spatial or spectral resolution, and as a result, improvement in classification abilities can be expected. Intuitively, we can consider such benefits as adding surface geometric information to the reflectance information, then more fine classifications can be conducted such as distinguishing between round roofs and triangle roofs.

2 Proposed Method

From Physical to Imaging Model

Physically, the hyperspectral imaging principle can be described using the bidirectional reflectance distribution function (BRDF) model. However, this model is rather complex for practice. In this part, we propose a more simplified yet efficient model that can be derived from the BRDF. We know that the BRDF can be described as:

$$f_r(x, y; \omega_i, \omega_r) = \frac{\mathbf{d}L_r(x, y; \omega_r)}{\mathbf{d}E_i(x, y; \omega_i)} \tag{1}$$

where L is the radiance on the location (x, y) and E is the irradiance on the location (x, y). The index i denotes the incident light, and the index r denotes the reflected light.

Equation (1) tells in detail the relations between the incident and the reflected light, from which we can see that the decisive point is the surface geometry and material properties. Although Eq. (1) is very accurate in describing the imaging processes, the integration calculations involved are too complex to solve in practice, and a more concise formula is preferred and needs to be explored. Here, we show how to use matrix analysis to derive a fine one.

For each pixel i, we can rewrite the imaging equation as:

$$I_i(\lambda) = E(\lambda)[\varrho_i(\lambda)]^{\gamma_i}\beta_i \tag{2}$$

where I is the spectral intensity at wavelength λ, E is the intensity of illumination at wavelength λ, ϱ is the reflectance of a single material at wavelength λ, γ is the material density of the target material, and β is the cos of the angle between the incident light and the surface normal.

If we rewrite Eq. (2) in the log-domain, we have:

$$\log I_i(\lambda) = \log E(\lambda) + \gamma_i \log \varrho_i(\lambda) + \log \beta_i \tag{3}$$

Hyperspectral images can deem as a 3D data cube; to present the data structure more clearly, we rewrite the 3D data as a 2D matrix, in which the rows are the pixels and the columns are the spectra. Then, the matrix expression of Eq. (3) is as follows:

$$\log\begin{bmatrix} I_1(\lambda_1) & \cdots & I_n(\lambda_1) \\ \vdots & & \vdots \\ I_1(\lambda_d) & \cdots & I_n(\lambda_d) \end{bmatrix} = \log\begin{bmatrix} E(\lambda_1) \\ E(\lambda_2) \\ \vdots \\ E(\lambda_d) \end{bmatrix} \mathbf{1}_n^{\mathrm{T}} + \log\begin{bmatrix} \rho_1(\lambda_1) & \cdots & \rho_n(\lambda_1) \\ \vdots & & \vdots \\ \rho_1(\lambda_d) & \cdots & \rho_n(\lambda_d) \end{bmatrix}$$
$$+ \mathbf{1}_d \log\left[\, \beta_1 \; \beta_2 \; \cdots \; \beta_n \,\right] \tag{4}$$

Since $\beta_i = L_x N_x^T + L_y N_y^T + L_z N_z^T$, subscribe this into Eq. (4). We can therefore simplify the imaging model of HSIs using the following expression:

$$\log(\boldsymbol{I}) = \log(\boldsymbol{E})\mathbf{1}_n^T + \log(\boldsymbol{\varrho}) + \mathbf{1}_d \log(L_x N_x^T + L_y N_y^T + L_z N_z^T) \tag{5}$$

Next we show how to use this model to obtain the reflectance and surface height from a single HSI. It is a two-part procedure. First, we apply the HSI decomposition method to extract reflectance alone with the shading component from the original HSI; then, we apply the shape from the shading method to obtain the surface normals from the shading component in procedure 1.

Reflectance Recovery
The cost function of our model is:

$$u = \left\| \log(\boldsymbol{\varrho}) - \log(\boldsymbol{\varrho})\boldsymbol{W}^{\mathrm{T}} \right\|_{\mathrm{F}}^2 + \left\| \log(\boldsymbol{\varrho}) + \log(\boldsymbol{E})\mathbf{1}_n^{\mathrm{T}} + \mathbf{1}_d \log(\boldsymbol{\beta})^{\mathrm{T}} - \log(\boldsymbol{I}) \right\|_{\mathrm{F}}^2 \tag{6}$$

The solution of Eq. (6) lies on the zero-point of the first partial derivative function. It is:

$$\begin{cases} \partial u/\partial \boldsymbol{\varrho} = \boldsymbol{\varrho}(\boldsymbol{I}_n - \boldsymbol{W}^{\mathrm{T}})(\boldsymbol{I}_n - \boldsymbol{W}^{\mathrm{T}})^{\mathrm{T}} + \boldsymbol{\varrho} + \boldsymbol{e}\mathbf{1}_n^{\mathrm{T}} + \mathbf{1}_d\boldsymbol{\beta}^{\mathrm{T}} - \boldsymbol{I} = 0 \\ \partial u/\partial \boldsymbol{e} = -(\boldsymbol{I} - \mathbf{1}_d\boldsymbol{\beta}^{\mathrm{T}} - \boldsymbol{\varrho})\mathbf{1}_n + \boldsymbol{e}\mathbf{1}_n^{\mathrm{T}}\mathbf{1}_n = 0 \\ \partial u/\partial \boldsymbol{\beta}^{\mathrm{T}} = -\mathbf{1}_d^{\mathrm{T}}(\boldsymbol{I} - \boldsymbol{e}\mathbf{1}_n^{\mathrm{T}} - \boldsymbol{\varrho}) + \mathbf{1}_d^{\mathrm{T}}\mathbf{1}_d\boldsymbol{\beta}^{\mathrm{T}} = 0 \end{cases} \tag{7}$$

Further, we can get:

$$\boldsymbol{\varrho} = \sigma(\mathbf{I}_d - \frac{1}{d}\mathbf{1}_d\mathbf{1}_d^T)I(\mathbf{I}_n - \frac{1}{n}\mathbf{1}_n\mathbf{1}_n^T)\boldsymbol{G}^{-1} \tag{8}$$

Once the reflectance is derived, the shading component can then be obtained using the following procedure:

$$\beta^T = \frac{1}{d}\mathbf{1}_d^T(I - \mathbf{e}\mathbf{1}_n^T - \rho) \tag{9}$$

$$= \frac{1}{d}\mathbf{1}_d^T(I - \frac{1}{n}(I - \mathbf{1}_d\boldsymbol{\beta}^T - \rho)\mathbf{1}_n\mathbf{1}_n^T - \rho) \tag{10}$$

$$= \frac{1}{d}\mathbf{1}_d^T I - \frac{1}{nd}\mathbf{1}_d^T I\mathbf{1}_n\mathbf{1}_n^T + \frac{1}{n}\beta^T\mathbf{1}_n\mathbf{1}_n^T + \frac{1}{nd}\mathbf{1}_d^T\rho\mathbf{1}_n\mathbf{1}_n^T - \frac{1}{d}\mathbf{1}_d^T\rho \tag{11}$$

$$= \frac{1}{d}\mathbf{1}_d^T I(\mathbf{I}_n - \frac{1}{n}\mathbf{1}_n\mathbf{1}_n^T) - \frac{1}{d}\mathbf{1}_d^T \rho(\mathbf{I}_n - \frac{1}{n}\mathbf{1}_n\mathbf{1}_n^T) + \frac{1}{n}\beta^T\mathbf{1}_n\mathbf{1}_n^T \quad (12)$$

$$= \frac{1}{d}\mathbf{1}_d^T \bar{I} - \frac{1}{d}\mathbf{1}_d^T \bar{\rho} + \frac{1}{n}\beta^T\mathbf{1}_n\mathbf{1}_n^T \quad (13)$$

Rewrite β^T in the matrix format. We then get the shading component $\boldsymbol{S}$, from which we can further recover surface normals.

Surface Normals

The shading component of HSIs can be presented by a $d \times n$ matrix, with d spectral band and n pixels. To estimate surface normals from shading is to decompose $\boldsymbol{S}$ into the product of one $d \times 3$ matrix and one $3 \times n$ matrix, which can be expressed by:

$$\boldsymbol{S} = \boldsymbol{LN} = \begin{bmatrix} L_{1a} & L_{1b} & L_{1c} \\ L_{2a} & L_{2b} & L_{2c} \\ \vdots & \vdots & \vdots \\ L_{3a} & L_{3b} & L_{3c} \end{bmatrix} \begin{bmatrix} N_{1a} & N_{2a} & \dots & N_{na} \\ N_{1b} & N_{2b} & \dots & N_{nb} \\ N_{1c} & N_{2c} & \dots & N_{nc} \end{bmatrix} \quad (14)$$

where $\boldsymbol{L}$ is a $d \times 3$ matrix denoting illumination, whereas $\boldsymbol{N}$ ia a $3 \times n$ matrix denoting surface normals.

Our goal is then to derive $\boldsymbol{L}$ and $\boldsymbol{N}$ satisfying such conditions. We first perform singular value decomposition (SVD) on $\boldsymbol{S}$:

$$\boldsymbol{S} = U \Lambda V^T \quad (15)$$

where U and V are orthogonal matrices and Λ is a diagonal matrix in which the bulk of energy is concentrated in the first several columns.

Here, we use the first three columns of Λ to approximate Λ:

$$\boldsymbol{S} \approx U\sqrt{\Lambda^{d3}}\sqrt{\Lambda^{3n}}V^T \quad (16)$$

where Λ^{d3} denotes the first three columns of Λ respectively, and Λ^{3n} denotes the first three rows of Λ.

Let $\boldsymbol{N} = \sqrt{\Lambda^{3n}}V^T$, and we obtain the surface normals; from the normals, we can further compute the surface height $H = H(x, y)$ of the image. We observe the following relation:

$$(n_a, n_b, n_c) = \frac{1}{\sqrt{h_a^2 + h_b^2 + 1}}(h_a, h_b, -1) \quad (17)$$

which leads to

$$\begin{cases} h_a = -\frac{n_a}{n_c} \\ h_b = -\frac{n_b}{n_c} \end{cases} \tag{18}$$

Since

$$h_a(a,b) = H(a+1,b) - H(a,b) \tag{19}$$

$$h_b(a,b) = H(a,b+1) - H(a,b) \tag{20}$$

Solving the following overconstrained linear systems leads to the solution of surface height:

$$\begin{cases} n_c H(a+1,b) - n_c H(a,b) = n_a \\ n_c H(a,b+1) - n_c H(a,b) = n_b \\ n_b(H(a,b) - H(a+1,b)) = n_a(H(a,b) - H(a,b+1)) \end{cases} \tag{21}$$

3 Experiments and Results

Data
MUUFL Gulfport is used in our experiments. The data were acquired by a CASI-1500 flown over the University of Southern Mississippi Gulf park Campus [5]. The resolution of the data is 325 × 337, with a spatial resolution of 1 m. There were 64 spectral bands of data (Fig. 1).

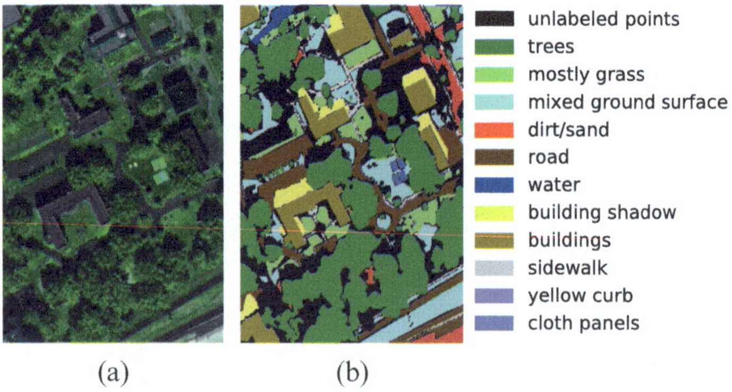

Fig. 1 MUUFL Gulfport data set. (**a**) False-color composite image. (**b**) Ground reference map

Results and Analysis

The reflectance and surface normals obtained by the proposed method are shown in Fig. 2. The reflectance component gives the intrinsic properties of a certain material and defines its response to the light under a certain wavelength. Thus, it is a function of wavelength, and in visual presentation, only three bands are selected and false color images are generated and provided. The surface normals component describes the geometric structures of the ground surface. It is very helpful to provide additional discriminating information regarding object recognition.

To verify the efficiency of our method, we implement the obtained reflectance and surface normal components as input features of support vector machines (SVMs), which constitute one of the most extensively used classifiers. As the reflectance component provides the spectral information, whereas the surface normals component provides the spatial information; thus, an improvement in classification performance is expected when two complementary forms of information are applied. The proposed method is compared with three alternative methods, which are:

1. Raw-SVM: the input features to the SVM are the original spectral information.
2. EMP-SVM: the input features to the SVM are the extended morphological profiles extracted from HSIs [6].
3. EMAP-SVM: the input features to the SVM are the extended morphological attribute profiles extracted from HSIs [7].

The classification maps generated using the proposed method and the alternative methods are shown in Fig. 3, from which we can see that the proposed method has the most satisfactory result in terms of boundary accuracy. The overall accuracy (OA) of each method is given in Table 1, from which we can see that the proposed method achieves the highest performance, especially when the training samples

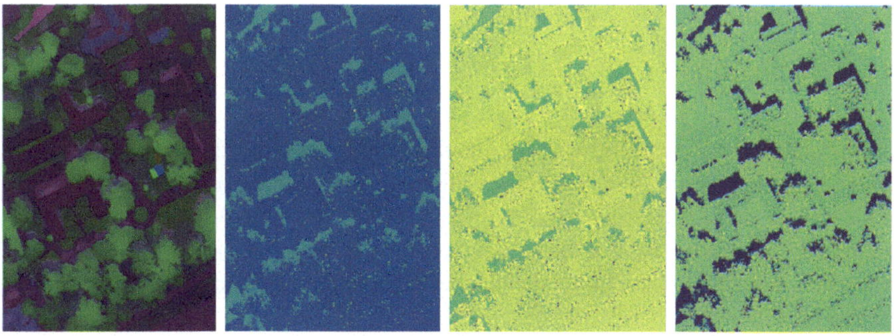

(a) (b) (c) (d)

Fig. 2 Recovery result: (**a**) reflectance, (**b**), (**c**), and (**d**): surface normals component, n_a, n_b and n_c

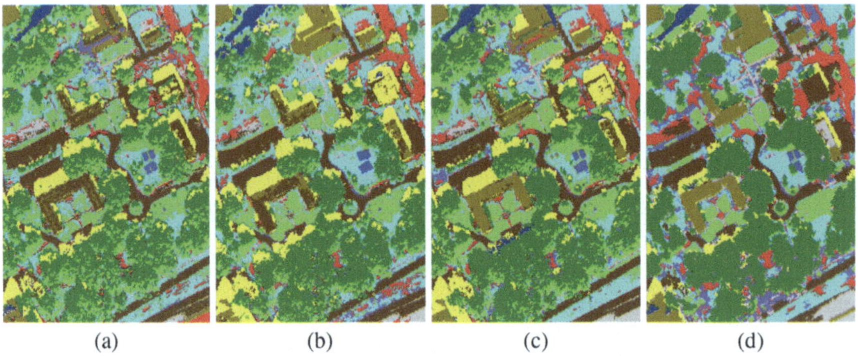

Fig. 3 Classification maps generated by (a) Raw-SVM, (b) EMP-SVM, (c) EMAP-SVM, and (d) the proposed method SVM, support vector machine; EMAP, extended morphological attribute profile

Table 1 Overall accuracy (%) of different methods

Method	5	10	15	20
Raw-SVM	65.82 ± 4.54	71.31 ± 2.82	73.76 ± 3.02	74.83 ± 2.81
EMP	68.44 ± 4.89	74.03 ± 2.48	77.47 ± 2.60	81.29 ± 1.61
EMAP	67.52 ± 5.19	76.68 ± 2.67	79.47 ± 2.15	81.93 ± 1.83
Proposed	**71.90 ± 3.98**	**77.81 ± 2.94**	**80.18 ± 2.01**	**83.00 ± 1.95**

The best results of the experiments are given with bold values

are very small. This demonstrates that the efficiency of physically meaningful geometric information is more helpful in classification than artificial morphological information.

4 Conclusion

In this chapter, we proposed a novel method for recovering reflectance and surface normals from a single HSI. Since the reflectance component provides the intrinsic information on a material, whereas the surface normals describe the surface geometry structure, we use these two complementary forms of information to improve classification performance. The experimental results demonstrate that our method is efficient.

Acknowledgements This work was supported in part by the National Science Fund for Excellent Young Scholars under Grant 61522107 and the National Natural Science Foundation of key international cooperation under the Grant 61720106002. This work was also supported by a key research and development project of the Ministry of Science and Technology (No.2017YFC1405100). (Corresponding author: Yanfeng Gu.)

References

1. Barron, J. T., & Malik, J. (2014). Shape, illumination, and reflectance from shading. *IEEE Transactions on Pattern Analysis and Machine Intelligence, 37*(8), 1670–1687.
2. Tappen, M. F., & Freeman, W. T. (2005). Recovering intrinsic images from a single image. *IEEE Transactions on Pattern Analysis and Machine Intelligence, 27*(9), 1459–1472.
3. Barron, J. T., & Malik, J. (2012). Shape, albedo, and illumination from a single image of an unknown object. In *IEEE Computer Society Conference on Computer Vision and Pattern Recognition* (pp. 334–341). Piscataway, NJ: IEEE.
4. Saxena, A., Sun, M., & Ng, A. Y. (2009). Make3d: Learning 3D scene structure from a single still image. *IEEE Transactions on Pattern Analysis and Machine Intelligence, 31*(5), 824–840.
5. Gader, P., Zare, A., Close, R., Aitken, J., & Tuell, G. (2013). *Muufl Gulfport Hyperspectral and Lidar Airborne Data Set*. Florida Technical Report: 570.
6. Benediktsson, J. A., & Pesaresi, M. (2001). A new approach for the morphological segmentation of high-resolution satellite imagery. *IEEE Transactions on Geosciences and Remote Sensing, 39*(2), 309–320.
7. Dalla Mura, M., Atli Benediktsson, J., Waske, B., & Bruzzone, L. (2010). Extended profiles with morphological attribute filters for the analysis of hyperspectral data. *IEEE Transactions on Geosciences and Remote Sensing, 31*(22), 5975–5991.

Characteristics Extraction of Behavior of Multiplayers in Video Football Game

Zhiwen Wang, Hao Ouyang, Canlong Zhang, Bowen Tang, Zhenghuan Hu, Xinliang Cao, Jing Feng, and Min Zha

Abstract In the process of behavior recognition of multiplayers for soccer game video, various features of athletes need to be extracted. In this paper, color moments extracted by using color classification learning set are regarded as color feature. Contour features of athletes are extracted by utilizing players silhouettes block extraction and normalization. Hough transform is used to extract the features of coordinates of pitch line, which can be used for camera calibration, rebuilding the stadium, and calculating the coordinate of players in the real scene. The trajectories of players and ball are predicted by using Kalman filter, while trajectories characteristics of player and ball are extracted by using the trajectory growth method. Temporal and spatial interest points are extracted in this paper. Experimental results show that the accuracy of behavior recognition can be greatly improved when these features extracted are used to recognize athlete behavior.

Keywords Behavior recognition · Feature extraction · Color moment · Hough transform · Trajectory growth method · Temporal and spatial interest points

1 Introduction

In order to detect whether each pixel represents a feature, the extraction of behavior characteristics of multiplayers in soccer video is a primary operation in the process

Z. Wang (✉) · H. Ouyang
College of Computer Science and Communication Engineering, Guangxi University of Science and Technology, LiuZhou, Guangxi Province, People's Republic of China

C. Zhang
College of Computer Science and Information Technology, Guangxi Normal University, Guilin, Guangxi Province, People's Republic of China

B. Tang · Z. Hu · X. Cao · J. Feng · M. Zha
College of Electrical and Information Engineering, Guangxi University of Science and Technology, Liuzhou, Guangxi Province, People's Republic of China

E. T. Quinto et al. (eds.), *The Proceedings of the International Conference on Sensing and Imaging, 2018*, Lecture Notes in Electrical Engineering 606,
https://doi.org/10.1007/978-3-030-30825-4_11

of behavior recognition [1]. The result of feature extraction is that the different subsets divided by the points on the image, which often belong to isolated points, continuous curves, or contiguous regions. Usually, there are many features that can be used for feature recognition behavior in soccer video game, such as the color feature of the stadium, ball, athletes and referees, the moving trajectory (spatial relationship) features of ball and athletes, the contour feature of players, the shape features of ball pitch line features, etc.

2 Related Work

On the feature extraction of athletes, in recent years, many researchers have carried out extensive research on it. Among them, the color features of the clothing of athletes, the contours features of the athletes, and the characteristics of the field line are extracted. In the past 10 years, the motion analysis based on contour features has developed into an important feature extraction technique. Recently, a lot of research has been done in this area. In [2, 3], novel approaches for indexing and evaluating sketch-based systems were proposed. A recent work in sports analytics [4] uses sketch-based search techniques to analyze player movements in rugby matches. It uses multiple distance-based similarity measures, which compares the user sketch against video scenes. Beside the traditional image and video search, sketch-based search techniques can also be applied to other complex data sets. In [5, 6], 2D sketches are used to search for 3D objects. Lee et al. [7] make use of synthesized background shadows to help users in sketching objects. Further applications of sketch-based search include retrieval in graphs [8], or for bivariate data patterns in scatter plots [9, 10]. Regarding data generation, in [11] 2D shape sketches are used to create and manipulate high-dimensional data spaces. A novel model-based technique that is capable of generating personalized full-body 3D avatars from orthogonal photographs [12]. The proposed method utilizes a statistical model of human 3D shape and a multiview statistical 2D shape model of its corresponding silhouettes.

3 Characteristics Extraction of Behavior of Multiplayers

Color Feature Extraction of Players and Referees

Since soccer is an ornamental sport, the playing field, ball pitch line, the clothing of referee and players are designed to have unique visual effects. From visual features, these color differences are one of the best messages [13–16]. The field is green and the field lines are white. Judges and players must wear as high a contrast as possible. Color characteristics can be used not only to improve tracking ability, but also to distinguish players which belong to different teams. Therefore, the color feature of video images can be extracted for behavior recognition [2]. In this paper, color classification learning sets is used to find interesting regions by mapping the image pixels to their color classes, and then morphological operators is used to group pixels. The mixed color space is used to detect the best distinction of the space produced by the pixels of the opponent team and the referee [17, 18].

Color Classification, Segmentation, and Feature Extraction

The first step in dealing with video images of a soccer game is to apply color classification and segmentation to the image. Assume that the previous color classification learning set includes the green field set, clothing set of team, and other set, color image pixels are firstly mapped to the respective classification in visual perception module, and then morphological operations is used to find the region of interest by denoising of the same type of color grouping region [3, 4].

In addition, the object of interest can be characterized by the characteristics of the image patch. In particular, the assumption of using the player's upright position to contact the pitch plane (if considered in a particularly long image sequence, at least) [19, 20]. The size of every patch can be estimated with much accuracy by using image patch coordinates. In addition, the interest objects must meet certain compact relations (the ratio between area and perimeter), such as the players and soccer. These assumptions can be used to filter data and more reliably extract related objects.

Existing color regions and color patches can be used to define the interest complex regions that are handled by certain image processing operations. For example, in order to find the line of field, the green centralized area in the field image (not including the player or referee occlusion area) are considered to get the field line. This area can be represented by using Eq. (1).

$$\mathrm{L} = (\theta \cap \mathrm{C}) - \alpha 1 - \alpha 2 - \alpha 3 \tag{1}$$

where, L is the area of the field line, θ is non-green area, C is playfield detection, $\alpha 1$ is the area where the team 1 is located, $\alpha 2$ is the area where the team 2 is located, and $\alpha 3$ is the area where the referee is located.

After obtaining the color regions of interest, the color moments of the region can be extracted as color features. In the HIS color space, the central moment (the first three order color moments) of each component can be calculated by Eq. (2).

$$\begin{cases} M_1 = \frac{1}{N}\sum\limits_{i=1}^{N} X(p_i) \\ M_2 = \left[\frac{1}{N}\sum\limits_{i=1}^{N} (X(p_i) - M_1)^2\right]^{1/2} \\ M_3 = \left[\frac{1}{N}\sum\limits_{i=1}^{N} (X(p_i) - M_1)^3\right]^{1/3} \end{cases} \quad (2)$$

Among them, X represents the H, I and S components in the HIS color space. $H(p_i)$ represents the X value of the ith pixel of the image p. N is the number of pixels in the image. Figure 1 is the image of the RGB distribution, the HIS color space, H, I, and S components and histogram. The calculated center moment of image and seven characteristic values are 1.0e + 004 * [1.5278 0.000 0.0001 0.0000 0.0004 0.0002 0.0003 0.0009 0.0005 0.0007].

Robustness of Color Classification

Lighting conditions change when the camera sweeps from one side of the court to another, clouds change, rain begins, and so on. For these reasons, reliable color segmentation cannot be achieved by prior learning of color categories and remains unchanged during the game [21–26]. On the contrary, color segmentation must adapt to the change of the sky, especially the segmentation of green sites in color categories. Therefore, the expectation maximization method is used to segment the green sites in color classes. By giving the stadium model and estimating camera parameters, it is determined that these areas must be green. The related area is expressed by Eq. (3).

$$\mathrm{A_R} = A_\mathrm{P} - A_\mathrm{T1} - A_\mathrm{T2} - A_\mathrm{Re} - A_\mathrm{L} \quad (3)$$

where, A_R is the related area, A_P is the playfield detection, A_T1 and A_T2 represent the area in which the team 1 and team 2 is located respectively, A_Re is the region of the referee, and A_L is the area where the pitch of a course line is located. Morphological operations are used to eliminate holes in the region in this paper. Then, the pixels in these regions are extracted to estimate the color classes of green regions. Finally, we use this color model to estimate camera parameters. In practice, the estimation of the classification model is much lower than that of the camera parameter estimation.

Contour Feature Extractions of Players

Given a video $v = \{I_1, I_2, \cdots, I_T\}$ which contains T frames of soccer game behavior, the related behavior contour sequence $S_s = \{s_1, s_2, \cdots, s_T\}$ can be obtained from

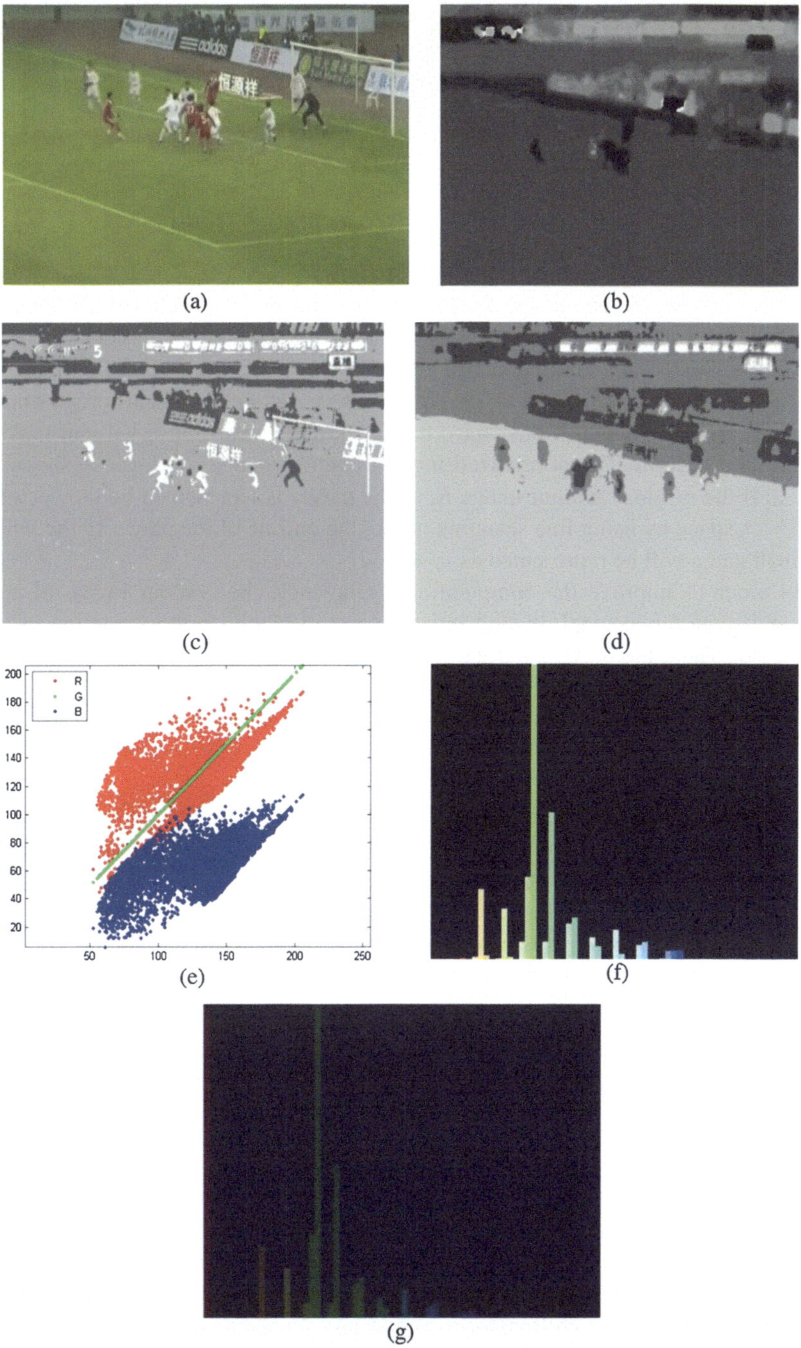

Fig. 1 Color component and histogram of video football match image. (**a**) Original image, (**b**) *H* component of image, (**c**) *I* component of image, (**d**) *S* component of image, (**e**) Color distribution of image, (**f**) *H*-*S* histogram of image, (**g**) *H*-*I* histogram of image

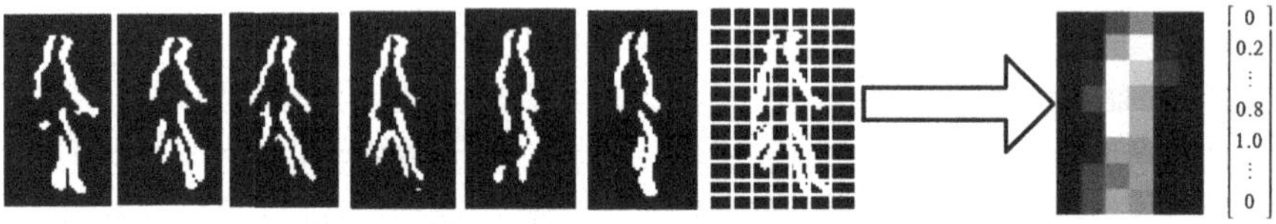

Fig. 2 Contour sequence and block feature representation of a player

the original video. The size and position of the foreground region changes with the distance between the player and the camera, the size of the target, and the behavior that have been completed changes. On the basis of keeping the contour ratio of the player unchanged, the contour image of the player is centralized and normalized, so that the resulting image $\mathrm{RI} = \{R_1, R_2, \cdots, R_T\}$ contains as much foreground as possible. All input video frames have the same dimension $r_i \times c_i$ without making the motion deform. The normalized motion contour image of player is shown in Fig. 2. If the original contour image R_i of the player is represented by the vector r_i in $\Re^{r_i \times c_i}$ space by using line scanning mode, the outline of the player in the whole football game will be represented as $v_r = \{r_1, r_2, \cdots, r_T\}$.

In order to improve the computational efficiency, the contour image of each player is equidimensional divided into $h \times w$ sub blocks without overlap in this paper [5]. Then, the normalized values of each sub block are calculated by using Eq. (4).

$$N_i = b(i)/mv, \quad i = 1, 2, \cdots, h \times w \tag{4}$$

Among them, $b(i)$ is the number of foreground pixels of the ith sub block and mv is the maximum of all $b(i)$. In $\Re^{h \times w}$ space, the descriptor of the silhouette of a player in the tth frame is $f_t = [N_1, N_2, \cdots, N_{h \times w}]^T$ and the outline of the player in the whole video can be correspondingly represented as $vf = \{f_1, f_2, \cdots, f_T\}$. In fact, the original player contour representation v_{r} can be considered as a special case based on block features, that is, the size of sub block is 1×1 (one pixel).

Feature Extractions for Stadium Line

The field line contains important coordinate information of the stadium. The results of line extraction can be directly used for camera calibration, stadium reconstruction, and calculation of the coordinates of players in the real scene. In the process of extracting court line feature, the video image of ball game is translated into binary image, then the feature of coordinate parameters of court line is extracted preliminarily using Hough transform, and at last the accurate linear coordinates is computed by using gray fitting. When using Hough transform to extract line feature, distance vector which can be calculated by Eq. (5) can be used to represent.

$$d = x\cos\theta + y\sin\theta \tag{5}$$

Among them, the value range of d is the diagonal length l of the video image, that is $d \in [-l, l]$, θ is the intersection angle between the vertical axis y and the level axis x and $\theta \in [0, \pi]$. x and y represent the two-dimensional coordinates of image pixels. We define the parameter space with the subscript d and θ using an array of integers k, and set the threshold th. When we use Hough transform to run statistical calculation, if $k > th$, the curve with the subscript d and θ is determined a straight line. Because the calculated value d may be negative, the subscript d is rewritten as $d + l$. The specific steps are as follows [6].

Step 1: construction and initialization of the lookup table of function *sin* and *cos*.
Step 2: for each non-background point (white dot) of a binary image, each value of d corresponding θ is computed by using Eq. (4) and the values of $d + l$ and $k + 1$ are meanwhile computed.
Step 3: when the values are greater than th, all subscripts d and θ corresponding to array k are found out, and then $d - l$ is calculated.
Step 4: because of the field line has not been refined in the binary image to be detected, there is a certain width. In the case of Hough transformation, it leads to several groups of similar d and θ for a straight line at the same time. If there is similar d and θ, only one group is retained.
Step 5: the coordinates of the two ends of the field line obtained by Hough transformation are (x_0, y_0) and (x_1, y_1).The mean gray value $Mean_{i,j}$ of the pixels passing through the line between the point i and the point j in the gray image are calculated by using $((x_0 + \sin\theta(f - m \times i), y_0 - \cos\theta(f - m \times i))$ and $((x_1 + \sin\theta(f - m \times j), y_1 - \cos\theta(f - m \times j))$, respectively. Where, f expresses the range of fitting and m is expressed as the fitting step length, $i, j \in \left[0, \frac{2f}{m}\right]$. When the calculated $\mathrm{Mean}_{i,j}$ is maximum, the determined line segment i', j' corresponding to maximum $\mathrm{Mean}_{i,j}$ can be considered as the optimal course line. Figure 3 is the characteristic of the field line determined by using Hough transform.

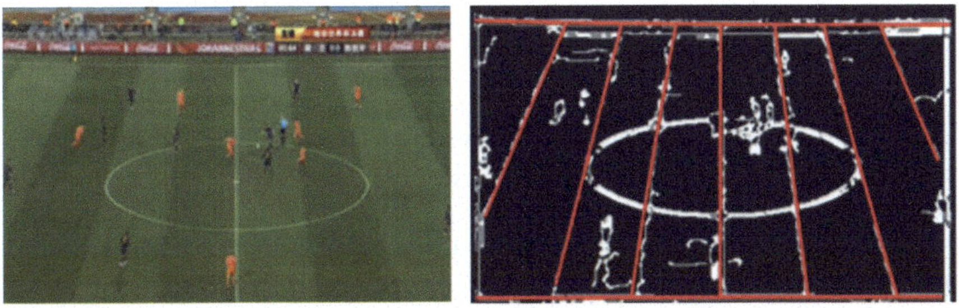

(a) (b)

Fig. 3 Features extraction of stadium line. (**a**) Original image, (**b**) Extracted field line

Feature Extractions for Motion Trajectory of Player and Ball

In the process of feature extraction for motion trajectory of players and ball, the video game is divided into small segments containing a specific number of video frames. When we extract the motion trajectory of players and ball, the segment is regarded as the basic unit, that is, the length of the processed trajectory is not more than the number of frames in the video fragment. After obtaining the candidate area of the motion target of each frame of soccer game video, first, we locate the moving objects in the three consecutive frames of the video in a spatial-temporal domain. The position of second frames in three consecutive frames is centered, and the candidate areas falling near the position are found in the front and back frames. After finding such a continuous three frame image, determine whether the moving target is included in the existing trajectory. If there is no, the new trajectory is initialized with the moving target in the three consecutive frame, and the location of each trajectory is recorded. After getting the new trajectory, the Kalman filter is used to predict the trajectory. The prediction Eq. (6) is used to predict new trajectory.

$$\begin{cases} X_t = AX_{t-1} + \gamma_t \\ O_t = BX_t + \kappa_t \end{cases} \tag{6}$$

Among them, $X_t = AX_{t-1} + \gamma_t$ is the motion equation of the system, $O_t = BX_t + \kappa_t$ is the observation equation of the system, X_t and O_t is system state vector and system state measurement vector of time *t*, respectively, γ_t and κ_t is the vector of the motion which is the normal distribution and measurement noise, and they are mutually independent. *A* and *B* express the state transfer matrix and the measurement matrix. The center position of the moving target is chosen as the measurement vector of the state of system, and the center position of the moving target and its motion speed are regarded as state vectors of the system. It can obtain Eq. (7) [7].

$$X = \begin{bmatrix} x \\ v_x \\ y \\ v_y \end{bmatrix}, \quad O = \begin{bmatrix} x \\ y \end{bmatrix}, \quad A = \begin{bmatrix} 1\,1\,0\,0 \\ 0\,1\,0\,0 \\ 0\,0\,1\,1 \\ 0\,0\,0\,1 \end{bmatrix}, \quad B = \begin{bmatrix} 1\,0\,0\,0 \\ 0\,0\,1\,0 \end{bmatrix} \tag{7}$$

Among them, (x, y) is the center position of the moving target, v_x and v_y express the motion speed of the moving target in the direction *x* and direction *y*. Prediction results by using Kalman filter are shown in Fig. 4, '+' is true value and 'o' is predictive value in the prediction process.

The location of the trajectory of the moving target in the new frame is predicted through Kalman filter, and the candidate moving targets near the location are searched in the frame image. If the motion target exists, the trajectories of the motion target are extended to the frame, and the center of the candidate moving target is used as the trajectory in the frame. If the corresponding candidate moving

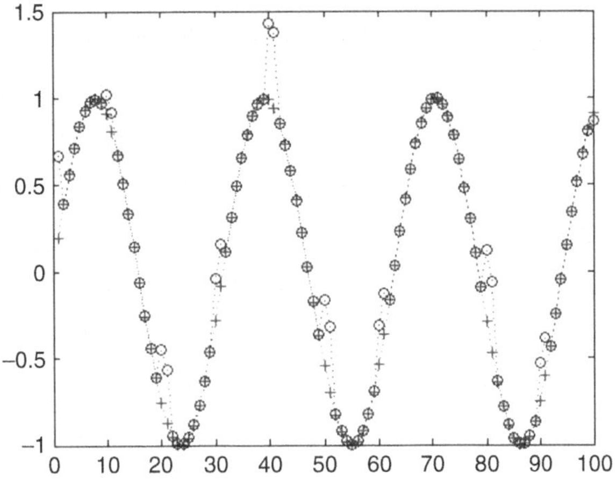

Fig. 4 Prediction process of Kalman filter

target is not found, the moving target corresponding to this trajectory is missing, blocked, or disappearing. When the false or occluded frames do not exceed the threshold, the trajectory of moving target is extended to this frame and the position of trajectory in this frame is instead of prediction value predicted by using Kalman filter. When the false or occluded frames exceed the threshold, we think this path has disappeared in the video, and then stop the growth trajectory. By means of trajectory growth, we can get a plurality of trajectory generated by candidate moving objects from the segments of soccer game video (including the moving target and noise), as shown in (a) in Fig. 5. The trajectory of red and blue is the track of two teams. Green trajectory is the track of soccer. When the soccer leaps and bounds in the air, its trajectory is mapped to the field. The reddish brown trajectory is the referee trajectory, and the red blue track is the track generated by noise. Because part of the generated trajectory is generated by noise, it is necessary to select the corresponding trajectories of the real moving target from these trajectories.

The set of trajectories generated by the real moving target is defined as C_{t}. The initialization set is C_{t}, the element in the set C_{t} is all the trajectories, that is $C_{\mathrm{t}} = \{T_i, i \in [1, N]\}$. Among them, T_i is the ith track of the video clip in the current football game. N is expressed as the number of the tracks in the video clip of the current football game. We randomly select two trajectories T_{u} and T_{v} with different starting frames in soccer video clips, $K_{\mathrm{s,\,u}}$, $K_{\mathrm{e,\,u}}$, $K_{\mathrm{s,\,v}}$, and $K_{\mathrm{e,\,v}}$ are corresponding to their starting frames and ending frames respectively and $K_{\mathrm{s,\,u}} \le K_{\mathrm{s,\,v}}$. When the end frame of the trajectory T_{u} is larger than the starting frame of the trajectory T_{v}, that is $K_{\mathrm{e,\,u}} \ge K_{\mathrm{s,\,v}}$, the trajectories T_{u} and T_{v} intersects in the space-time domain, $T_{\mathrm{u}} \cap T \neq \phi$. On the other hand, the two trajectories are considered to be separated. In the video segment of a football game, the trajectories of the moving targets are usually long, and the track produced by the noise is shorter. Therefore, when the two trajectories cross, the trajectory of the moving target takes a longer trajectory.

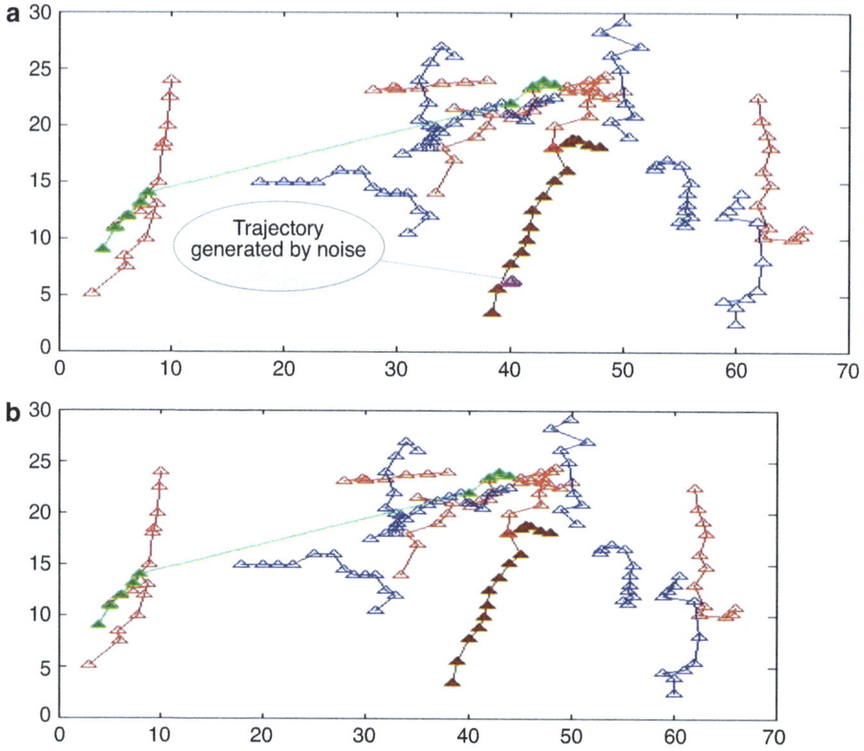

Fig. 5 Trajectory generation diagram for video football game fragment. (**a**) Trajectories of moving objects with noise trajectories. (**b**) Trajectories of moving objects without noise trajectories

We can use the Eq. (8) to calculate the set of trajectories generated by the moving target.

$$C_t = \begin{cases} C_t - \{T_u\} \text{ if } \left(\left(K_{e,u} - K_{s,u}\right) < \left(K_{e,v} - K_{s,v}\right)\right) \wedge (T_u \cap T_v \neq \phi) \\ C_t - \{T_v\} \text{ if } \left(\left(K_{e,u} - K_{s,u}\right) \geq \left(K_{e,v} - K_{s,v}\right)\right) \wedge (T_u \cap T_v \neq \phi) \end{cases} \tag{8}$$

By choosing the trajectory, the set C_t of the trajectories of the moving target is finally obtained, as shown in (b) in Fig. 5. The separate trajectories are included in set C_t, that is, the missing frames exist between the trajectories. The main reason for the missed frames is the mutual occlusion between the moving targets on the field and the sudden change in the direction and speed of the motion. In order to get the full trajectories of the video clip of the football game, it is necessary to connect these separate tracks.

First, the predicted values $\hat{p}_{k,u}$ and $\hat{p}_{k,v}$ of the trajectories T_u and T_v in the interval $[K_{e,u}, K_{s,v}]$ are calculated. Then we search for two points closest to the distance between the two trajectories in the predicted interval $[K_{e,u}, K_{s,v}]$, corresponding to the frame a and frame b in the trajectories T_{u} and T_{v}. In the prediction process, Eq. (9) is used as constraint condition.

$$\begin{cases} (a,b) = \arg\min\limits_{a,b} dist\left(\hat{p}_{a,\mathrm{u}}, \hat{p}_{b,\mathrm{v}}\right) \\ a \le b, \\ K_{\mathrm{s,u}} \le a \le K_{\mathrm{e,v}}, \\ K_{\mathrm{s,u}} \le b \le K_{\mathrm{e,v}}. \end{cases} \tag{9}$$

When $a \geq b$, the location of the moving target which is missed before the frame a and the location of the moving target is not detected after the frame a were expressed by the predicted value of trajectory T_{u} and track T_{v} in this interval, respectively. Their mean is used to represent the position of moving target in the frame a. We can use Eq. (10) to calculate this mean.

$$p_k = \begin{cases} \hat{p}_{k,\mathrm{u}} & K_{\mathrm{e,u}} \le k < a \\ \left(\hat{p}_{k,\mathrm{u}} + \hat{p}_{k,\mathrm{v}}\right)/2 & k = a \\ \hat{p}_{k,\mathrm{v}} & a < k \le K_{\mathrm{s,v}} \end{cases} \tag{10}$$

When $a < b$, the location of the missed moving target around frame a can be calculated as same as the time $a \geq b$. The motion of the moving target is smaller in the two frames a and b, and the more accurate position of the moving target is obtained by using the linear interpolation method (Eq. 11).

$$p_k = \begin{cases} \hat{p}_{k,\mathrm{u}} & K_{\mathrm{e,u}} \le k \le a \\ (k-a)\left(\hat{p}_{b,\mathrm{v}} - \hat{p}_{a,\mathrm{u}}\right)/(b-a) & a < k < b \\ \hat{p}_{k,\mathrm{v}} & b \le k \le K_{\mathrm{s,v}} \end{cases} \tag{11}$$

The location of the missed moving target can be accurately filled through the connection, and the complete track of moving target can be generated.

Extraction of Temporal and Spatial Interest Points of Players and the Referee

The temporal and spatial interest points refer to the relatively large intensity changes in time and space. Temporal and spatial interest point representation is a relatively new underlying representation method of characteristics for frame sequences. The video sequence of football game video is set as $v : \Re^2 \times \Re \rightarrow \Re$. Linear scale space representation $R : \Re^2 \times \Re \times \Re_+^2 \mapsto \Re$ is constructed by convolution with anisotropic Gaussian kernel with different spatial and temporal variances σ_{r}^2 and τ_{r}^2 of v. The relationship between them can be expressed by using Eq. (12).

$$R\left(\cdot;\sigma_{\mathrm{r}}^{2},\tau_{\mathrm{r}}^{2}\right)=g\left(.;\sigma_{\mathrm{r}}^{2},\tau_{\mathrm{r}}^{2}\right)*v\left(\cdot\right) \tag{12}$$

The space-time separable Gauss kernel is defined by Eq. (13).

$$g\left(x,y,t;\sigma_{\mathrm{r}}^{2},\tau_{\mathrm{r}}^{2}\right)=\frac{\exp\left(-\left(x^{2}+y^{2}\right)/2\sigma_{\mathrm{r}}^{2}-t^{2}/2\tau_{\mathrm{r}}^{2}\right)}{\sqrt{(2\pi)^{3}\sigma_{\mathrm{r}}^{4}\tau_{\mathrm{r}}^{2}}} \tag{13}$$

It is of great concern to use separate scale parameters for both temporal and spatial domains, since the temporal and spatial components of events are generally independent. Moreover, the detection of events by temporal and spatial interest point operators depends on the observed time and space scales, so it is necessary to deal with the corresponding temporal and spatial scale parameters σ_{r}^{2} and τ_{r}^{2} separately.

In this paper, we consider a 3×3 space-time second-order moment matrix consisting of first-order space and time scales, and Gauss weighting function is used to the average of Gauss kernel by Eq. (14).

$$g\left(.;\sigma_{\mathrm{r}}^{2},\tau_{\mathrm{r}}^{2}\right)\mu=g\left(.;\sigma_{\mathrm{r}}^{2},\tau_{\mathrm{r}}^{2}\right)*\begin{pmatrix} {r_x}^2 & r_xr_y & r_xr_t \\ r_xr_y & {r_y}^2 & r_yr_t \\ r_xr_t & r_yr_t & {r_t}^2 \end{pmatrix} \tag{14}$$

The scale parameters σ_{i}^{2} and τ_{i}^{2} in Eq. (14) are fused into the local scale parameters σ_{r}^{2} and τ_{r}^{2} by using $\sigma_{i}^{2}=s\sigma_{\mathrm{r}}^{2}$ and $\tau_{i}^{2}=s\tau_{\mathrm{r}}^{2}$, the first derivative is defined by Eq. (15).

$$\begin{cases} r_x\left(x,y,t;\sigma_{\mathrm{r}}^{2},\tau_{\mathrm{r}}^{2}\right)=\partial x\left(g*v\right) \\ r_y\left(x,y,t;\sigma_{\mathrm{r}}^{2},\tau_{\mathrm{r}}^{2}\right)=\partial y\left(g*v\right) \\ r_t\left(x,y,t;\sigma_{\mathrm{r}}^{2},\tau_{\mathrm{r}}^{2}\right)=\partial t\left(g*v\right) \end{cases} \tag{15}$$

In order to detect interest temporal and spatial points, we search for the regions with significant eigenvalues λ_1, λ_2 and λ_3 in μ in video. In different methods of searching region, spatial domain Harris corner function is defined as Harris corner function in space-time domain by combining determinant and tracking extension of μ in Eq. (16).

$$H=\lambda_1\lambda_2\lambda_3-k(\lambda_1+\lambda_2+\lambda_3)^3,\quad (\lambda_1\leq\lambda_2\leq\lambda_3) \tag{16}$$

In order to show that the positive local maximum of H corresponds to the high value point λ_1, λ_2, and λ_3, the ratio is defined as $\alpha=\lambda_2/\lambda_1$ and $\beta=\lambda_3/\lambda_1$, and the Eq. (16) can be rewritten as Eq. (17).

$$H=\lambda_1^3(\alpha\beta-k\left(1+\alpha+\beta\right) \tag{17}$$

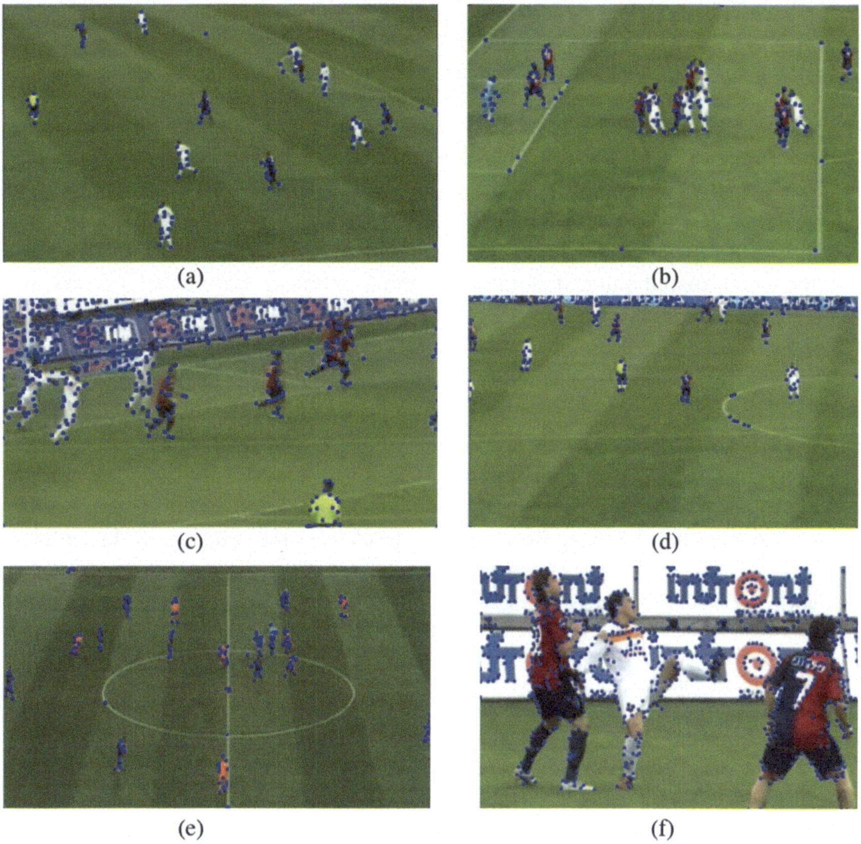

Fig. 6 Detection results of temporal and spatial interest points. (**a**) 52nd frame, (**b**) 88th frame, (**c**) 137th frame, (**d**) 223rd frame, (**e**) 76th frame (another video), (**f**) 361st frame (close-up frames)

If $H \geq 0$, then $k \leq \alpha\beta/(1 + \alpha + \beta)^3$. Assume $\alpha = \beta = 1$, the maximum possible value of k is 1/27.When the value k is large enough, the point corresponding to the positive local maximum H varies sharply along the time and space directions. Especially when the maximum value of α and β in the spatial domain is 23, $k \approx 0.005$. Therefore, we can detect the temporal and spatial interest points in the video sequence v of football matches by detecting the positive local temporal and spatial maximum of H. Time and space interest point detection results are shown in Fig. 6.

4 Conclusion

Because single feature is difficult to effectively describe the behavior characteristics of multiple athletes, it is not reliable to use single feature to identify the behavior of multiple athletes. According to the characteristics of various features of athletes needed to extract in the process of multiplayers behavior recognition in soccer game video and feature extraction has a great influence on the final recognition results, in this paper, we extract the contour features, clothing color histogram, temporal and spatial interest points, and color moment features of players and referees. We also transform the soccer video images into binary images and extract feature of coordinate parameters of field line using the Hough transform, and calculate accurate linear coordinates using gray fitting. We use Kalman filter to track the moving object and predict its motion trajectory. In order to reduce the burden of high-level recognition algorithm, we propose using the method of growth trajectories to extract low-level features, such as trajectory features of the moving object. The experimental results show that it can greatly improve the accuracy of the behavior recognition by using these features to identify the behavior of the athletes.

Acknowledgments The authors are very grateful for the support provided by the National Natural Science Foundation of China (61462008, 61751213, 61866004), the Key projects of Guangxi Natural Science Foundation (2018GXNSFDA294001, 2018GXNSFDA281009), the Natural Science Foundation of Guangxi (2017GXNSFAA198365), 2015 Innovation Team Project of Guangxi University of Science and Technology (gxkjdx201504), Scientific Research and Technology Development Project of Liuzhou (2016C050205).

References

1. Lazebnik, S., & Raginsky, M. (2009). Supervised learning of quantizer codebooks by information loss minimization. *IEEE Transactions on Pattern Analysis and Machine Intelligence, 31*(7), 1294–1309.
2. Cao, Y., Wang, H., Wang, C., Li, Z., Zhang, L., & Zhang, L. (2010). Mindfinder: Interactive sketch-based image search on millions of images. In *Proceedings of the international conference on multimedia* (pp. 1605–1608). New York: ACM.
3. Eitz, M., Hildebrand, K., Boubekeur, T., & Alexa, M. (2011). Sketch-based image retrieval: Benchmark and bag-of-features descriptors. *IEEE Transactions on Visualization and Computer Graphics, 17*(11), 1624–1636.
4. Legg, P. A., Chung, D. H. S., Parry, M. L., Bown, R., Jones, M. W., Griffiths, I. W., & Chen, M. (2013). Transformation of an uncertain video search pipeline to a sketch-based visual analytics loop. *IEEE Transactions on Visualization and Computer Graphics, 19*(12), 2109–2118.
5. Eitz, M., Richter, R., Boubekeur, T., Hildebrand, K., & Alexa, M. (2012). Sketch-based shape retrieval. *ACM Transactions on Graphics, 31*(4), 31:1–31:10.
6. Lee, J., & Funkhouser, T. (2008). Sketch-based search and composition of 3D models. In *EUROGRAPHICS workshop on sketch-based interfaces and modeling*, June 2008.
7. Lee, Y. J., Zitnick, C. L., & Cohen, M. F. (2011). Shadowdraw: Real-time user guidance for freehand drawing. In *ACM SIGGRAPH 2011 papers* (pp. 27:1–27:10). New York: ACM.

8. von Landesberger, T., Bremm, S., Bernard, J., & Schreck, T. (2010). Smart query definition for content-based search in large sets of graphs. In *Proceedings of the International Symposium on visual analytics science and technology* (pp. 7–12). Geneva: Eurographics Association.
9. Scherer, M., Bernard, J., & Schreck, T. (2011). Retrieval and exploratory search in multivariate research data repositories using regressional features. In *Proceedings of the 11th Annual International ACM/IEEE joint conference on digital libraries* (pp. 363–372).
10. Shao, L., Behrisch, M., Schreck, T., von Landesberger, T., Scherer, M., Bremm, S., & Keim, D. A. (2014). Guided sketching for visual search and exploration in large scatter plot spaces. In *Proceedings of the EuroVA International workshop on visual analytics*. Geneva: The Eurographics Association.
11. Wang, B., Ruchikachorn, P., & Mueller, K. (Dec 2013). Sketchpadn-d: Wydiwyg sculpting and editing in high-dimensional space. *IEEE Transactions on Visualization and Computer Graphics, 19*(12), 2060–2069.
12. Michael, N., & Lanitis, A. (2014). Model-based generation of realistic 3D full body avatars from uncalibrated multi-view photographs. In L. Iliadis, I. Maglogiannis, & H. Papadopoulos (Eds.), *Artificial intelligence applications and innovations. AIAI 2014. IFIP advances in information and communication technology* (Vol. 436). Berlin: Springer.
13. Zhiwen, W., & Shaozi, L. I. (2014). Adaptive fractal-wavelet image denoising based on multivariate statistical model. *Chinese Journal of Computer, 37*(6), 1380–1389.
14. Ashok, V., Amit, R. C., & Rama, K. C. (2005). Matching shape sequences in video with applications in human movement analysis. *IEEE Transactions on Pattern Analysis and Machine Intelligence, 27*(12), 1896–1909.
15. Wang, L., & Yung Nelson, H. C. (2010). Extraction of moving objects from their background based on multiple adaptive thresholds and boundary evaluation. *IEEE Transactions on Intelligent Transportation Systems, 11*(1), 40–51.
16. Michael, N., Drakou, M., & Lanitis, A. (2017). Model-based generation of personalized full-body 3D avatars from uncalibrated multi-view photographs. *Multimedia Tools and Applications, 76*(12), 14169–14195.
17. Fuchs, Z. E., Casbeer, D. W., & Garcia, E. (2016). Singular analysis of a multi-agent, turn-constrained, defensive game. In *2016 American Control Conference (ACC) Boston Marriott Copley Place*, July 6-8, 2016. Boston, MA, USA (pp. 4705–4712).
18. Lin, S., Dominik, S., Benjamin, N., Manuel, S., & Tobias, S. (2016). Visual-interactive search for soccer trajectories to identify interesting game situations. *Electronic Imaging, Visualization and Data Analysis*, 510.1–510.10.
19. Boiman, O., & Irani, M. (2007). Detecting irregularities in images and in video. *International Journal of Computer Vision, 74*(1), 17–31.
20. Shintani, T., Nobata, M., Muneyasu, M. (2016). A-15-14 a Silhouette extraction method for moving objects based on image characteristics. In *IEICE Engineering Sciences Society/Nolta Society Conference*, 2016.
21. Ahmed, H. A., Rashid, T. A., & Sadiq, A. T. (2016). Face behavior recognition through support vector machines. *International Journal of Advanced Computer Science and Applications, 7*(1), 101–108.
22. Alba-Cabrera, E., & Godoy-Calderon, S. (2016). Generating synthetic test matrices as a benchmark for the computational behavior of typical testor-finding algorithms. *Pattern Recognition Letters, 80*(1), 46–51.
23. Heng, F., Jun, X., Yong, D., & Jinhai, X. (2016). Behavior recognition of human based on deep learning. *Geomatics and Information Science of Wuhan University, 41*(4), 492–497.
24. Kim, H., Lee, S., Kim, Y., Lee, S., & Lee, D. (2016). Weighted joint-based human behavior recognition algorithm using only depth information for low-cost intelligent video-surveillance system. *Expert Systems with Applications, 45*(C), 131–141.
25. Jiang, Q. (2016). Research of multiple-instance learning for target recognition and tracking. *EURASIP Journal on Embedded Systems, 2016*(1), 1–6.
26. Wang, X., Gao, B., Masnou, S., Chen, L., & Theurkauff, I. (2016). Active colloids segmentation and tracking. *Pattern Recognition, 60*, 177–188.

A Differential Resonant Edge Detection Sensor

Nathan Ida

Abstract The detection of metal edges is a common problem in manufacturing of metal products including strips and tapes, as well as in alignment of metal sheets for assembly of products. The detection may be as simple as detecting the position of the edge—either qualitatively, that is, the very existence of a metal edge or may be highly quantitative, whereby the exact position of the edge is sought for the purpose of alignment or assembly. Metal strips may be very thin or relatively thick, may be ferromagnetic or nonmagnetic and in some applications, such as during production, may be at high temperatures. The present work proposes a self-compensating, highly sensitive resonant sensor suitable for the industrial environment. The method relies on two split coils one for each edge, detecting the position of one edge of the strip. To compensate for a variety of effects such as jitter, drift, and frequency locking, the two coils are switched at a fixed rate onto a single resonator. The difference in resonant frequency of the two coils is calculated and is used to locate the edges and to generate correction signals for the purpose of positioning the metal strip. The results presented here are based on tests made on a scale prototype sensor.

Keywords Magnetic induction · Edge detection · LC oscillators · Frequency mixing · Resonance · Induction sensors

1 Introduction

There are a few basic methods that have been investigated for the purpose of edge detection, all relying on induction and the effect the metal strip has on the induced magnetic field. Perhaps the most obvious method is that of a loosely coupled transformer [1]. In this approach, a primary coil generates a magnetic field at an appropriate frequency (a relatively low frequency of the order of

N. Ida (✉)
Department of Electrical and Computer Engineering, The University of Akron, Akron, OH, USA
e-mail: ida@uakron.edu

E. T. Quinto et al. (eds.), *The Proceedings of the International Conference on Sensing and Imaging, 2018*, Lecture Notes in Electrical Engineering 606,
https://doi.org/10.1007/978-3-030-30825-4_12

10–50 kHz) and induces a voltage in a secondary coil. Any metal between the two coils changes the coupling between them and hence the induced voltage in the secondary coil. The method is particularly effective with ferromagnetic materials but requires the coils to be in relatively cross-proximity for reasonable sensitivity. The output of the secondary coil is then correlated with the position of the metal strip. To detect the strip edges or to control the strip position, two transformers are typically used, one for each edge of the strip. This arrangement also allows for differential measurements in which a center position produces zero output. In practical applications, the output of the system is compared with a calibration curve and the position of the edge is determined from this comparison. Sensitivity is much higher for ferromagnetic materials than for nonferromagnetic materials. The coils are shielded from outside influences by a ferromagnetic shield. A variation of the transformer method relies on the shielding effect of the strip on the induced voltage [2]. The transformer arrangement is essentially the same as the previous method except that a reference sensing transformer is used to sense the center of the strip and, since it is insensitive to the edges, it measures the minimum coupling. The outputs of the edge transformers are then compared with the output of this reference transformer to affect positioning of the strip. A third method, often associated with metal detection is the pulse induction method [3–5]. Although it is primarily used for metal detection in the classical sense, it can be used to detect strip edges as well. The method consists of driving a coil with a short pulse which induces a transient current in the metal. The coil is then disconnected from the source and connected to a detector to detect the induced voltage in the coil due to the residual field in the metal. The pulse induction method is common in commercial metal detectors but it can also be used in the industrial environment. It should also be mentioned here that strip edges can also be detected using capacitive methods as well as ultrasound and optical methods. In some cases, even radars have been adapted for this purpose.

The induction methods described above suffer from one basic drawback—low sensitivity. Because the separation between the coils of the transformer is relatively large and the coupling between the coils is low, the induced voltages are small resulting in low sensitivity of detection. A secondary issue is the measurement of the small changes in induced voltages which is often analog and prone to noise, requiring additional signal processing and filtering. The main attraction of induction methods is that they are simple, sturdy, and can operate in difficult environment in the presence of contaminants and at high temperatures.

2 A Resonant Method for Metal Strip Edge Detection

To improve sensitivity and simplify measurements, the present work proposes a resonant method of edge detection. The fundamental principle is to form resonant circuits whose resonant frequency depends on the amount of metal within the resonator (and hence on the position of the strip). Since resonance can be affected at much higher frequencies than is possible with induction methods, the change in resonant frequency can be significant resulting in higher sensitivity. In addition, measurement of frequency is very simple and noise can be eliminated by digital means. In essence the measurement is digital, simplifying both the sensor and the process and interfacing directly with digital controllers. Although the measurement of frequency is digital, analog outputs are also possible using frequency to voltage (*F*/*V*) converters if required.

3 General

The basic idea of this sensor is a pair of split, balanced coils, one half of each below and one half above the strip as shown in Fig. 1. Under balanced conditions, that is, when the metal strip is centered about the coils, the two resonators, consisting of the coils and a capacitor in an LC oscillator, would resonate at the same frequency. The two signals can be digitized and fed into a digital mixer (Fig. 2). The signal out of the mixer, which is the difference between the two frequencies, is proportional to the position of the metal strip about the center of the sensor. Assuming that the two oscillators operate at the same frequency f when the strip is centered between the two coils the output is null. A misaligned metal strip will produce a change in frequency Δf in each oscillator. One oscillator will increase in frequency whereas the other will decrease. The effect at the output of the mixer is a frequency equal to $2\Delta f$, which can be either measured directly using a microprocessor or can be fed into an *f to V* converter to produce an analog signal proportional to the misalignment of the metal sheet. The latter can be used as a feedback signal for a centering mechanism if the strip is to be kept centered.

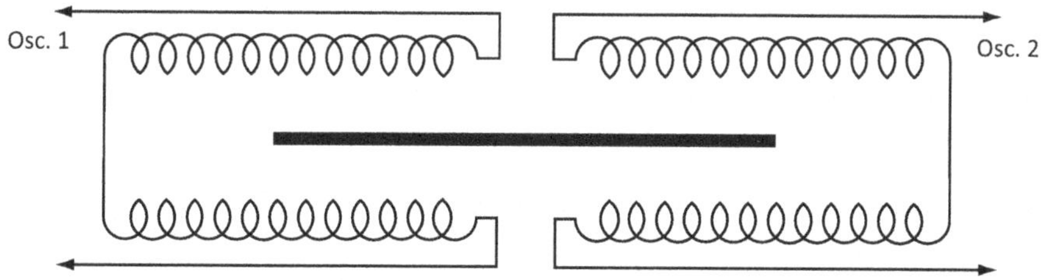

Fig. 1 Principle of the resonant sensor. Each coil is split in two equal parts connected in series

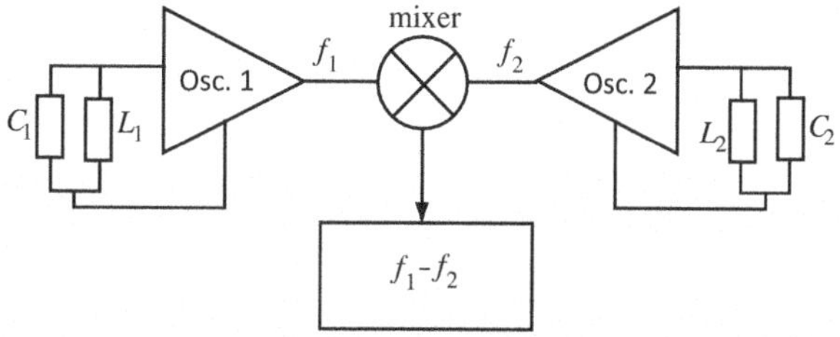

Fig. 2 Mixer method used to obtain the difference between the frequencies of two oscillators, either analog or digital

The configuration of coils in Fig. 1 has a number of advantages. First, there is a reduced sensitivity to the vertical position of the coils. This is important with strips that cannot be guaranteed to be perfectly flat when passing trough the resonator or if the strip does not coincide with the middle plane between the coils. Second, the split parts of the coils can be located at a sufficient vertical distance apart to allow space for the strips and to reduce heating effects from the strips. Perhaps the most important feature is the differential signal obtained from this configuration that allows for the reduction of common-mode effects such as temperature, external influences, and drift in properties of the coils.

Figure 2 shows the basic method of measurement of the differential frequency and is similar to the method of beat-frequency in common metal detectors. Depending on the type of oscillators used, the mixer can be analog (sinusoidal oscillators) or digital (pulse oscillators).

The mixing method in Fig. 2 has a few rather important disadvantages.

1. Frequency locking—This is a known problem when two oscillators operate at close frequencies and are in close proximities so that there is some coupling between them. Two oscillators, running at two different frequencies will lock to one another's frequency and change in tandem if the difference between the two frequencies is small and if there is cross-feedback between them. With the two coils in this sensor physically close to each other, this is inevitable. A simple solution is for the two oscillators to operate at two different frequencies so that when balanced (centered strip), the output of the mixer is a few kilohertz (a frequency difference of about 1–5% is sufficient to avoid locking). This is not a big problem since a constant frequency difference can be taken into account and subtracted either electronically or in the controlling microprocessor.
2. Digital mixer's jitter—Digital mixers rely on essentially dividing the incoming frequency. This produces outputs with varying pulse widths, which frequency counters interpret as change in frequency. Although these changes in frequency are usually small, so are the changes in frequency anticipated from this sensor. The normal method of eliminating this problem is to average a number of readings since the jitter tends to have zero average. At worst, this may slow the output of the sensor but it should not prove to be a big problem. In particular, when the output frequency is high, as required to avoid frequency locking, frequency jitter is usually negligible.
3. Digital mixer's frequency range—Digital mixers are cyclical. That is, if the output for f_1 and f_2 is Δf, then the output for nf_1 and mf_2 (n, m any integer) is still Δf. This is a minor point in that the difference between frequencies cannot be larger than the smaller of the two frequencies. Of course, the normal output of the sensor will only change in a range of a few percent of the basic frequency but it does mean that the frequency differential in (1) must be smaller or equal to $f_1/2$ (where f_1 is the lower oscillator frequency). Again, with a base frequency of one oscillator of, say 250 kHz, the second oscillator should be no higher than 375 kHz or no lower than 125 kHz. This is usually a sufficient range of operation.

4. Oscillator drift—Regardless of the type of oscillator in use, the frequency varies with time and external conditions. There are many effects that cause this variation including temperature and environmental effects on components, ageing of components and the like. The two oscillators are very likely to experience slightly different conditions, again introducing an error in the output. This error is more or less random but it can be significant. There are of course methods of stabilization and compensation of some of the effects that cause drift but these cannot entirely eliminate it, especially with the large coils and long wires needed in this application.

LC oscillators will drift in frequency for a variety of reasons, some which are easier to eliminate than others. There are of course the obvious reasons including changes in temperature and their effects (changes in resistance, capacitance, inductance, etc.), and other environmental effects. Then there are effects due to variations in power supply voltage, internal changes in components due to ageing and due to operating voltages and currents. For example, most capacitors will vary in capacitance with temperature and with current through them. Some of these effects are cumulative, some are not. Other influences that exist are due to electrostatic effects, external noise, etc. The net effect of all this is that the frequency can vary by a few percent over hours or days unless methods of stabilizing the frequency are applied.

First, of course, components must be selected properly, particularly capacitors. The coils must to be electrostatically shielded (common practice in electromagnetic devices), power sources require proper regulation and, perhaps, the whole circuit, minus the external coils may need to be temperature stabilized and electromagnetically shielded. The latter method is common in high-stability circuits.

There are also a number of methods of stabilizing the frequency through feedback—essentially looking at changes in frequency and changing the capacitance through use of varactor diodes but this only works well for fast changes in frequency. Slow changes such as changes due to slow drift of the metal strip to one side or the other are likely to be missed by a method of this type.

4 The Proposed Method

To avoid the need for complex compensation circuits needed to eliminate some of the difficulties above, the present work adopts a simple and effective approach that can minimize the errors without the expense and complexities implied above. The solution is to use a single oscillator and switch the two coils alternately so that the only component that changes is the coil (see Fig. 3). With the left coil connected, the resonant frequency of the left resonator is first obtained, the left coil is switched out and the right coil switched in and its resonant frequency measured. The process is repeated indefinitely. The two frequencies are measured at small intervals (10–40 ms) and compared continuously. Since all components and conditions are the

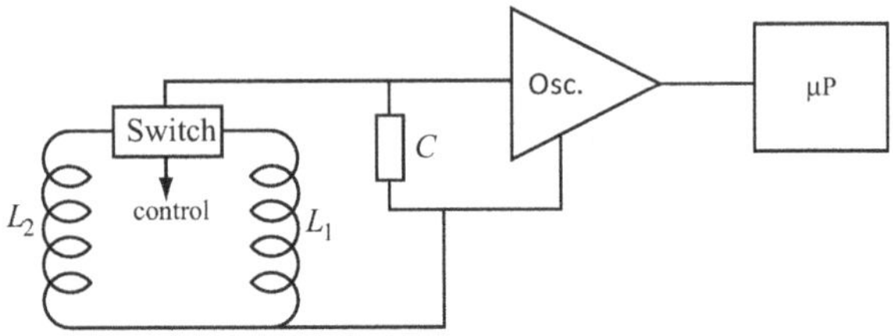

Fig. 3 Switched coil oscillator

same, (except for the coils) most of the problems associated with drift and other common-mode effects are eliminated or reduced significantly. The approach also eliminates any possibility of frequency locking. Once calibrated this gives much more stable output and there is less of a need of using sophisticated stabilization circuits. In experiments performed with this approach using manual switching of the coils, the maximum drift between two subsequent measurements was below 5 Hz. Faster switching can reduce this drift to entirely negligible values.

5 Oscillators

A number of oscillators were used in a prototype, with preference to square wave oscillators for the ease of experimentation. The oscillator ranged from very simple gate delay oscillators (Fig. 4a) to Colpitts type oscillators in a number of configurations (two are shown in Fig. 4b, c). In principle, it does not matter what type of oscillator is used as long as it can be made relatively stable and because of the switching of the coil, relatively fast settling. This issue turned out to be the main problem with LC oscillators and, in fact with this method of sensing as a whole. The switching in of the coil means that the time constants must be short and requires a low inductance coil and, to ensure noise free switching, low current through the inductor. To do that the coil is made as two rectangular multi-turn printed circuit coils in a spiral. The construction of one half of a coil is shown in Fig. 5. Each pair is connected in series to affect the resonator on each side of the strip. To reduce the coils' resistance, and hence increase the Q-factor of the resonator, the coil is printed on both sides of the PCB with the traces on the two sides in parallel. The coil used in the prototype is 60 cm by 24 cm with 22 cm separation between the two parts of each coil, an inductance of 170 μH, and a resistance of 0.2 Ω. The two halves of the coil for each side are separated 22 cm (vertical distance in Fig. 2) with a minimum gap between the left and right coils of 5 cm (for the smallest strips). This gap ensured minimal coupling between the left and right coil when used to sense narrow strips but can be much larger for wider strips. The coils are positioned so that the strip, when centered, runs with its edges close to the center of the coils in Fig. 2. This ensures that for small deviation of the strip, the response is approximately linear since in the vicinity of the center of the coils the field is more or less uniform.

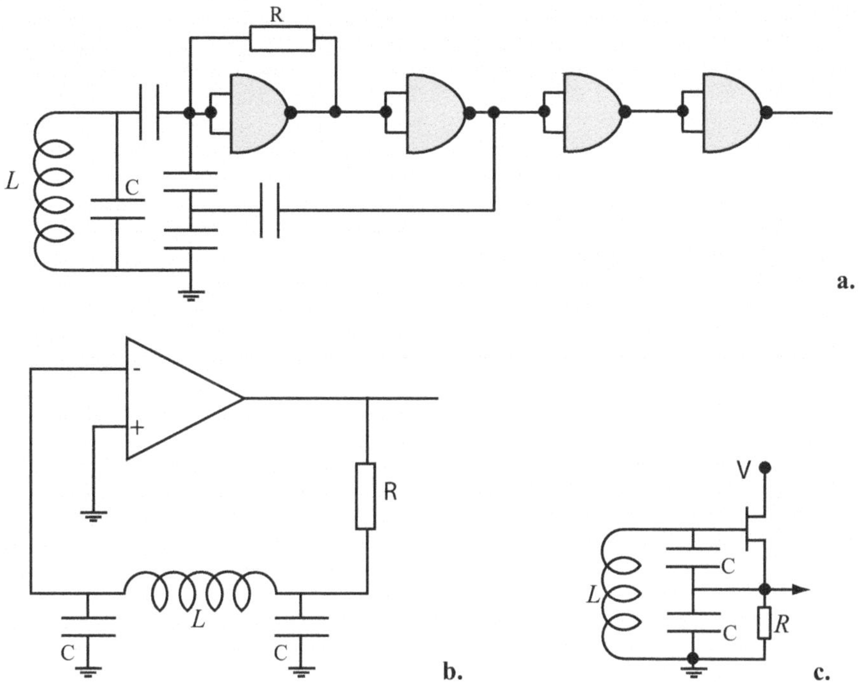

Fig. 4 Simplified schematics of oscillators used for testing. (**a**) A gate delay LC oscillator. (**b**) An Op-amp based Colpitts oscillator. (**c**) An FET Colpitts oscillator

6 Selection of Operating Frequency

As a rule, the higher the frequency, the more sensitive the sensor should be since the change in frequency due to the position of the metal sheet will be larger.

In the experiments performed on the oscillators in Fig. 4, the frequency ranged from about 100 kHz to about 3 MHz. Higher frequencies can be used of course but much more care must be taken in the implementation of the oscillator and prototyping becomes much more difficult. A frequency of about 250 kHz (when no strip is present) proved to be the most convenient for experimentation. This is accomplished by ensuring that the total capacitance connected across the inductor is approximately 2.2 nF.

7 Calibration

Calibration of the sensor is rather trivial. First, the output is zeroed by adjusting the position of the strip. Then a relation between the position of the strip can be obtained by simple measurements. However, in practice none of that is required: all that is

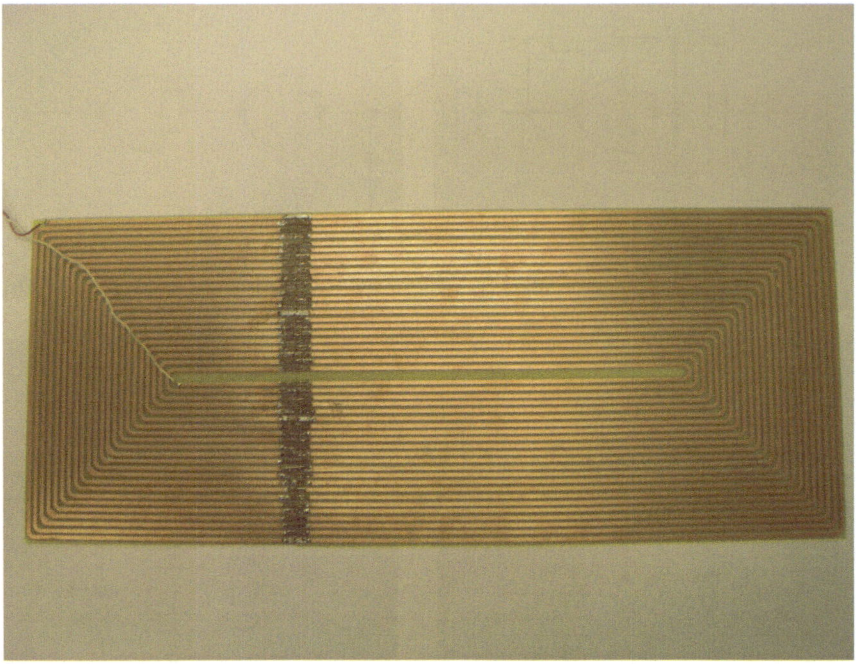

Fig. 5 Structure of the coil. The coil was made in two sections and the traces soldered together to form the complete coil. Each side of the sensor consists of two such coils connected in series

needed is to generate a correcting signal with the sole purpose of restoring the zero output which then ensures that the strip is centered.

8 Switches

The switches needed to switch the coils are analog CMOS switches available commercially. Since the needs of this application are very modest, almost any switch can be used. The important parameters are the low on resistance but even more importantly, high off resistance to ensure minimal coupling between the two coils through the switches. A single pole double throw switch capable of switching at a rate of 100 to 200 times per second and 250 kHz through signal is sufficient.

9 Operation of the Sensor

Operation consists of the following steps:

Connect left coil to the oscillator for 25 ms.
Wait 10 ms for stabilization of the oscillator.

Turn on a counter fed by a 100 MHz crystal oscillator.
Count for 2048 pulses of the input.
Turn off the counter.
Disconnect the left coil.

A. Read the last output of the counter.

Calculate the frequency and store in microprocessor.
Wait 5 ms.
Connect right coil to the oscillator for 25 ms.
Wait 10 ms for stabilization.
Turn on the counter fed by a 100 MHz crystal oscillator.
Count 2048 pulses of the input.
Turn off the counter.
Disconnect the right coil.

B. Read the output of the counter.

Calculate the frequency and store in microprocessor.
Compare the results in A and B using the microprocessor (or through a series of XOR gates).
Wait 5 ms.
Repeat the process indefinitely.

The process is illustrated in Fig. 6. The timings given here are based on the prototype described above and a base resonant frequency of the order of 250 kHz.

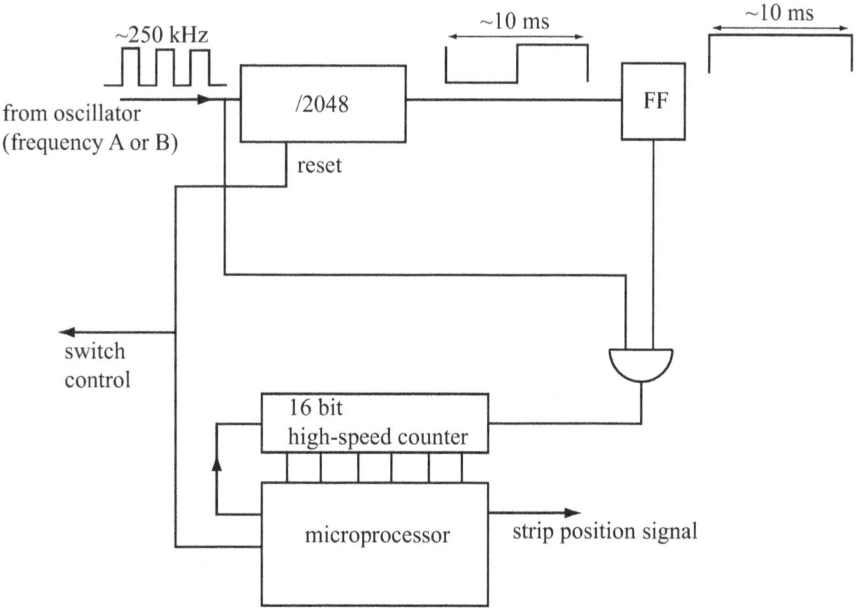

Fig. 6 Signal flow for the sensor

10 Performance

Although the sensor described in the previous paragraph is relatively simple its performance is quite good.

Sensitivity With the split coils at 22 cm apart, the sensitivity is 45 Hz/mm side movement of a 1.6 mm aluminum strip. This rises to 263 Hz/mm for a 1 mm steel strip.

Linearity Linearity depends on coil separation and on the distance of the coils from each other. For a vertical separation of 22 cm and 18 cm horizontal separation (strip is 48 cm wide), the output is linear for 36 mm side movement. Beyond that a slight nonlinearity can be observed that may require a calibration curve. Linearity improves with reduction in vertical separation but the horizontal separation has little effect as long as the centers of the coils are above the edges of the strip.

Deadband As was mentioned above, the issue of drift cannot be entirely eliminated, that is, the oscillators will drift slightly between two measurements. Therefore the uncertainty in the shift of the resonant frequency is about 5 Hz. That is, any shift of 5 Hz or less cannot be attributed to a side motion of the strip. In the minimum sensitivity of 45 Hz/mm, 5 Hz corresponds to about ± 0.12 mm. This must be considered a deadband or error in positioning.

Resolution The error due to drift also defines the resolution of the sensor. Thus, the lowest resolution is 0.12 mm as well. The resolution increases with reduction in coil separation and depends on the material as well.

Errors In addition to the errors due to drift, there are also operational errors that may come into play. Unlike the induction sensors described in the introduction, the resonant sensor can be affected by dielectric materials as well as conducting media. That is, introduction of any conducting material will affect the resonant frequency and any object that is not symmetric to both coils will change the output. Fixed objects such as shields, both symmetric and nonsymmetric can be taken into consideration through calibration but moving objects or objects that have not been compensated for will introduce errors. The sensor must be both electrostatically and magnetically shielded for best performance.

Performance parameters such as sensitivity and resolution can be improved by increasing the resonant frequency but much higher frequencies are not advisable unless the inductance of the coil can be reduced. This can be done by decreasing the number of turns or reducing the dimensions of the coil. This however reduces the uniformity of the fields between the coils and hence reduces linearity. The resolution can be improved by reducing the switching times for measurement of the resonant frequency. Switching times are dictated mostly by the settling time of the oscillator and there is very little room to reduce that.

11 Conclusions

The sensor described in this work relies on the shift in resonant frequency due to the presence of conducting and magnetic materials in an LC resonator in which only the inductor is influenced by the object under test. The use of a differential scheme of measurement reduces the effects of many common-mode conditions such as the effects of temperature on the resonant frequency. The differential method also simplified the design in that circuits are simple and components non-critical. The sensitivity is better than that of induction systems available commercially while other issues such as the need for shielding are comparable to induction methods.

References

1. Munker, W., Schwan. H., & Wagner, H. (2003). *Method and device for measuring the position of a metal strip*, US patent 6,580,262.
2. Wagner, H. (1994). *Methods and apparatus for measuring the position of an edge of a metal strip by electronic induction*, US Patent 5,339,029.
3. Kim, B., Yoon, J. W., Lee, S. E., Han, S. H., & Kim, K. (2015). Pulse induction metal detector with time-domain bucking circuit for landmine detection. *Electronics Letters, 51*(2), 159–161.
4. Rerkratn, A., Petchmaneelumka, W., Kongkauroptham, J., Kraisoda, K., & Kaewpoonsuk A. (2011). Pulse Induction Metal Detector Using Sample and Hold Method, 2011. In *11th International Conference on Control, Automation and System*, October 26–29, 2011. Gyeonggi-do Korea: IEEE Xplore.
5. Medek, R., Nicolics, J., & Schrottmayer, D. (2001). High sensitive pulse inductive eddy current measurement for mine detection systems. In *24th International Spring Seminar on Electronics, Calimanesti-Caciulata, Romania*, May 2001 (pp. 207–211). https://doi.org/10.1109/ISSE.2001.931058.

Microwave and Millimeter Wave Sensors for Nondestructive Testing and Evaluation

Nathan Ida

Abstract Microwave nondestructive testing and the sensors required to affect the testing are less well known than other electromagnetic methods such as eddy currents techniques. Nevertheless, they occupy a critical role in testing as well as imaging of conditions and flaws in primarily nonmetallic structures including plastics, polymers, and ceramics as well as wood structures, various composites and concrete. It has also found uses in conductors for surface testing and for evaluations of coatings. The methods and sensors range from microscopy to industrial scale testing and from simple open-ended waveguides to complex radars. The frequency range covers the whole microwave range and down to mm waves. The present work discusses the use of microwave testing, its place in a nondestructive testing regime, and a sampling of sensors that are representative of the art and its present state, with emphasis on emerging methods.

Keywords Microwave sensors · mm-wave sensors · Nondestructive testing · Microwave microscopy · Resonant testing methods · Radar

1 Introduction

Microwave nondestructive testing (NDT) is a method of testing materials and structures that uses microwaves and millimeter waves as the source of energy. The method can be viewed, to some extent, as an extension and adaptation of methods that have been developed for other purposes including radar, communications, and even microwave heating. But there are also techniques that have evolved specifically for NDT.

Historically, microwaves use for NDT started shortly after its first use in radar in attempts to quantify moisture and detect foreign objects in dielectrics [1]. Soon

N. Ida (✉)
Department of Electrical and Computer Engineering, The University of Akron, Akron, OH, USA
e-mail: ida@uakron.edu

E. T. Quinto et al. (eds.), *The Proceedings of the International Conference on Sensing and Imaging, 2018*, Lecture Notes in Electrical Engineering 606,
https://doi.org/10.1007/978-3-030-30825-4_13

it became an accepted method of testing of a variety of conditions and flaws in dielectrics, ceramics, and composite materials with applications to composite materials [2–4] and for testing of ceramics [5]. The main impetus for development of microwave methods of testing came in the 1970s with the increased availability of commercial microwave equipment and in particular the appearance of the network analyzer.

Testing with microwaves is defined and limited by properties of materials in the microwave field and the frequencies used for testing. The range of microwaves is formally defined between 300 MHz and 300 GHz. This corresponds to wavelengths between 1 m at 300 MHz and 1 mm at 300 GHz. Beyond this frequency range, formally it is the terahertz region and the methods of testing expected are different, having more to do with infrared radiation than classical electromagnetic waves. The range of microwaves defines the fundamental spatial resolutions that can be expected based on the wavelength. This resolution defines the smallest artifacts that can be discerned. However, there are many techniques that can be brought to bear to increase this resolution many fold. These include reliance on phase changes, which can vary rapidly, the use of evanescent waves, which decay within a fraction of a wavelength, testing with resonant structures, etc. Dimensional parameters such as thickness can be measured to well within a micrometer and scanning methods can easily improve resolutions. Methods borrowed from radar but on a much smaller scale, including SAR are also used for that purpose. Other methods include interferometric and microscopy methods which are not limited by the wavelength resolution. Microwave testing relies on the effects material properties have on electromagnetic waves. These include primarily effects of permittivity, conductivity, and permeability and any effect, property, or quantity that affects one or more of these. In addition, electromagnetic waves are affected by dimensional properties including volume and surface artifacts and interfaces. This makes for a large number of possible test responses. In lossless and lossy dielectrics, material composition, density and porosity, uniformity, delamination between layers and constituents, moisture content, contamination, etc. can be tested. In conductors, electromagnetic waves can only penetrate to very shallow depths and hence tests are limited to surface conditions such as roughness, scaling, and surface corrosion as well as surface-breaking flaws and dimensional variations of conductors. Unlike most testing methods, microwaves are also sensitive to properties of gases and hence can be used to detect and quantify the presence and density of gases including those emitted by explosives. Thus, the range of materials in which microwave NDT can be used is rather large and includes most plastics and polymers, ceramics, composite materials, concrete, biological and organic materials, many conducting media, wood and food products and, in fact, almost any medium in which electromagnetic waves of proper frequency can interact with its constituents.

A unique feature of microwave NDT is the fact that energy is usually coupled into the test medium through air and there is no need for contact. This is done through antennas of various types, open waveguides, and in some cases directly through the fields of transmission lines. Testing can also be done in microwave cavities or apertures in cavities can be used for this purpose. Testing is often done in the near fields of the sources because of the high gradient in the fields in that region but can also be done in the far field. Of course, like any other method of testing, microwave NDT has its problems. Signals can be noisy because of the many effects that influence it and results can be ambiguous for the same reasons. Reflections from structures near and far can affect tests and interference from a variety of electromagnetic sources can reduce sensitivity.

Microwave NDT relies on the interaction of high-frequency electromagnetic waves with materials, taking advantage of absorption in materials, scattering, attenuation, reflection, transmission, and resonance. These are fundamental effects and testing attempts to use these in various arrangements for quantitative evaluation of properties and conditions in materials.

There are many methods through which microwaves can be used for testing. The first that come to mind are those associated with propagation of electromagnetic waves and the reflection, transmission, and scattering of these waves due to the presence of flaws or material variations [6–9]. The same principles have been used for dimensional testing [10, 11]. Radar and radar-like methods are obvious applications that rely on reflection and scattering of electromagnetic waves. A particularly useful method of NDT is the ground penetrating radar (GPR) ([12–15]), but also related methods including newly developed ultra wide band (UWB) radars for low power, short range testing in dielectrics [16–23] and in biological media medical applications such as detection of tumors [24–27].

Particularly useful are tests in which the resonant frequency of resonant structures such as cavity resonators or resonant transmission lines are affected by the effect that is measured. These include moisture content monitoring and evaluation of material properties [28–33]. Resonant methods are particularly useful when they apply because their sensitivity and resolution can be orders of magnitudes better than, for example, reflection methods and these are not related to the wavelength.

Another method that has been successfully used to circumvent the wavelength limitation is the use of open-ended waveguides. These near-field methods are often referred to as microwave microscopy to indicate their ability to resolve well below the wavelength. (See for example: [32, 34–41].)

As is well known, microwave testing is applicable mostly in dielectrics and low loss dielectrics including ceramics, plastics, and polymers [3, 42, 43]. Lossy materials such as concrete, composites, carbon fibers based composites and sands can also be tested successfully [7, 44]. Another area of successful use of microwaves is in biological media, including medical applications [9, 25, 26].

Material properties can be tested for moisture content [29–31], consistency, and porosity [3, 6, 43–45] and other properties including density changes.

One of the earliest uses of microwave was for dimensional measurements [33, 46–49]. Some of these tests allowed resolutions in the micrometer range [42, 43]. Other applications are in the detection of disbands, delamination, and corrosion [2, 35], as well as surface conditions on conductors [50].

In addition, microwaves can also be used to detect surface conditions such as roughness, evaluate coatings on metals and, to a limited extent, detect fatigue cracks.

The number of applications of microwaves for testing and evaluation is quite large but, in effect, these applications can be summarized to that of detection and evaluation of changes in the dielectric constant, in permeability and in conductivity and, naturally, anything else that affects the electric properties of the material.

The following sections expand on a few of the more advanced methods of microwave testing and evaluation including microwave microscopy, resonant methods, GPR, and UWB radar.

2 Microwave Microscopy

Microwave microscopy relies on the behavior of open waveguides or apertures in cavity resonators. These are near-field methods that take advantage of the fact that under these conditions the gradients in the fields are very large although they are of very short extent, that is, the fields of apertures or open waveguides can only be used over a very short distance. The resolution of these methods stems from the very high spatial frequency they afford [32, 36, 51]. This high spatial frequency is also at the root of the method's limitations—it requires scanning for the detection of features of any extent, it must be done in very close proximity to the effects, and necessarily only small samples or sections of structures can be tested. The method is usually associated with localized testing and very often used for evaluation of material properties [37, 52, 53], rather than for testing for defects, although it has been used for dimensional testing and flaws in conductors [40, 54].

Microwave microscopy instrumentation is relatively simple. In its simplest form it consists of an open waveguide or transmission line (Fig. 1a), sometimes supplemented by a sharp tip that helps in increasing spatial resolution around the tip (Fig. 2a). Better resolution is obtained by using a small aperture in a resonant cavity or a resonant probe as shown in Figs. 1b and 2b. The fields around the aperture or tip may be viewed as being generated by an antenna in the near field and are reflected and scattered by the sample's surface and any feature or condition in the test medium. The signal obtained from the non-resonant probe is either from the scattered or reflected field. Because the fields are very low, it is only practical to

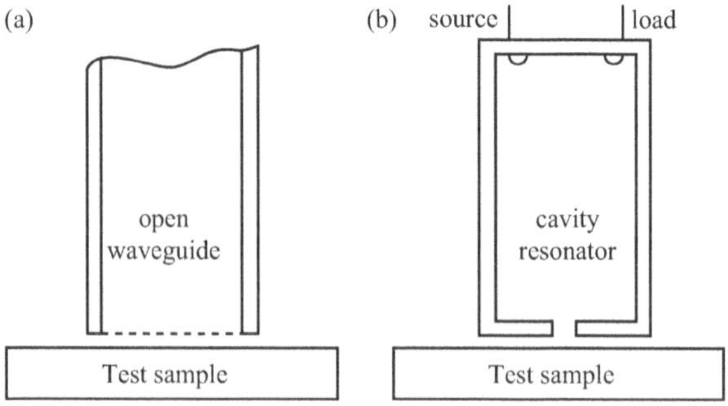

Fig. 1 Simple microwave microscopy probes. (**a**) Open waveguide. (**b**) Small aperture in cavity resonator

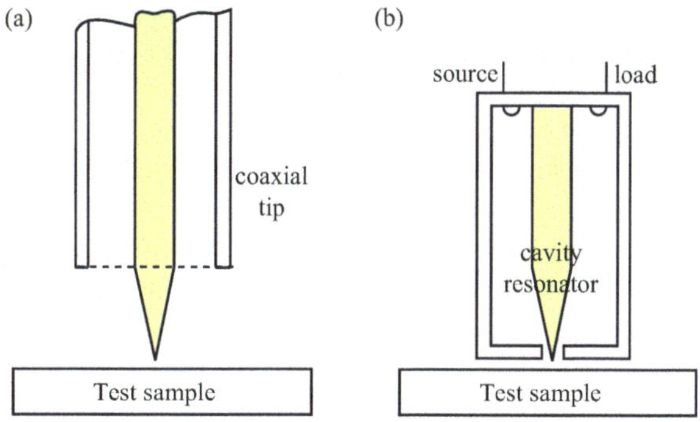

Fig. 2 Sensitive microwave microscopy probes (**a**) Tip in open coaxial probe. (**b**) Tip in cavity resonator aperture probe

monitor the impedance of the probe and extract the changes in material properties (complex permittivity or complex permeability) from the near-field impedance of the probe. The latter is approximated as [36]:

$$Z_s \approx \frac{1}{j\omega\varepsilon_0\varepsilon_s v} \tag{1}$$

Any deviation from this reference impedance is viewed as being due to change in material properties.

More sensitive measurements are affected through use of a resonant probe such as that in Fig. 1b or 2b. In these cases it is not the impedance that is monitored but, rather, the resonant frequency. In essence, the cavity is perturbed by the external loading due to the material properties and its resonant frequency changes due to the external conditions. This shift in resonant frequency may be written as [30]

$$\frac{f-f_0}{f} = -\frac{\int_v \Delta\varepsilon \mathbf{E}\cdot\mathbf{E_0}^* dv + \int_v \Delta\mu \mathbf{H}\cdot\mathbf{H_0}^* dv}{\int_v \mu\mathbf{H}\cdot\mathbf{H_0}^* dv + \int_v \varepsilon\mathbf{E_0}^*\cdot\mathbf{E}\, dv} \tag{2}$$

E and **H** in Eq. (2) are the perturbed fields, $\mathbf{E_0}$, $\mathbf{H_0}$ are the unperturbed fields, and $\Delta\varepsilon$, $\Delta\mu$ are the effective changes in the permittivity and permeability of the cavity caused by the external effects. This expression is only a simplistic representation but it serves to explain the behavior. In some applications including thin films, the expression in Eq. (2) can be simplified, especially if the relative permeability is 1 [30, 36]:

$$\frac{f-f_0}{f} \approx -(\varepsilon_{rf}-1)\frac{\varepsilon_0\int_{v_f} \Delta\varepsilon \mathbf{E}\cdot\mathbf{E_0}^* dv}{\int_v \mu_0 H_0^2 dv + \int_v \varepsilon_0 E_0^2 dv} \tag{3}$$

where v_f and ε_{rf} are the volume and permittivity of the film under the influence of the aperture, whereas the denominator is the total energy stored in the cavity. Similar expressions may be obtained for testing on conducting surfaces and other conditions.

Microwave microscopy was first developed for material characterization in thin films using ferromagnetic resonance [36] but has been adapted to application in NDT [37, 55, Zoughi 2000].

A simple demonstration of the possibility of microwave testing can be found in [37, 56]. In this application, the permeability of a thin film ferromagnetic CO-NETIC alloy is tested with the probe in Fig. 2b under the effect of an external magnetizing field and without the field. The basic configuration is shown in Fig. 3. The probe is stationary but the sample is scanned under the probe and the frequency response of the probe is recorded using a network analyzer. The microwave probe operated in reflection mode [55] and the frequency response was obtained from the S_{11} parameter. The test configuration was also simulated using commercial computational electromagnetic software.

As an example of the type of results can be obtained, consider Fig. 4. The figure compares numerical and experimental resonance data for an alloy sample. In this case the relative permeability is 1 and no external magnetic field is applied [56]. The numerical data was obtained using the transmission line modeling method (TLM) as described in [34]. Although relatively lengthy, modeling of microwave microscopy is quite practical. The simulation and experiment were repeated after application of an external magnetic field on a sample with relative permeability of 1000. The net effect of the higher permeability material is to shift the resonant frequency by approximately 180 kHz.

In another test, the values of the S_{11} parameter obtained at 889.61 MHz was used to sense and quantify changes in relative permittivity by scanning across a small 2-mm sample of relative permittivity of 1000. The results are shown in Fig. 5.

Although the result in Fig. 5 is excellent as is the comparison with the numerical simulation, the sample is very small and thin and the scan was done very slowly to

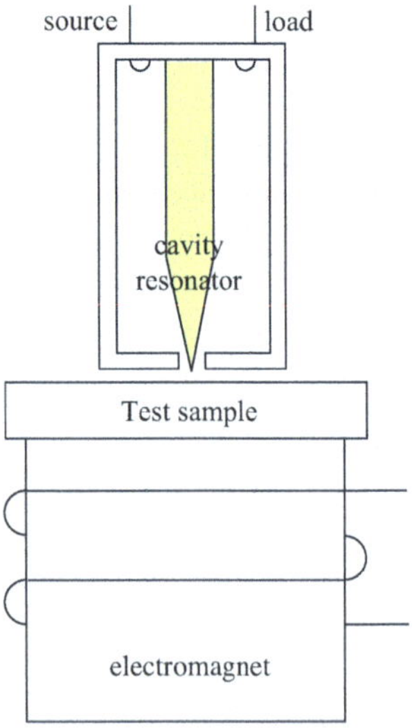

Fig. 3 Block-diagram of the experimental set-up: The sample moves during the scan whereas the probe is stationary. The electromagnet is used to bias the sample with a given magnetic flux density

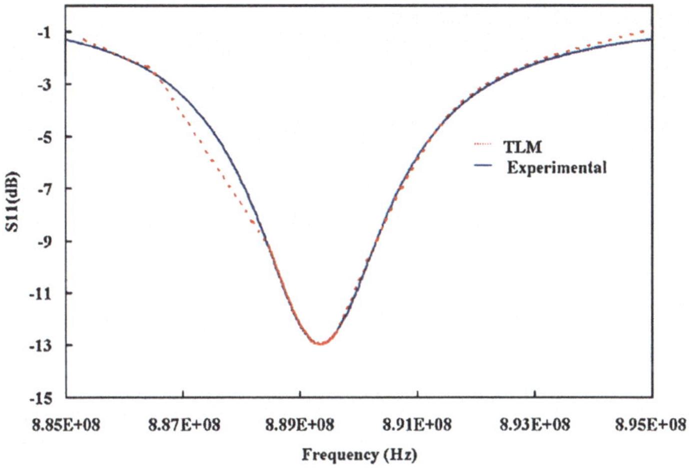

Fig. 4 Comparison between numerical (dashed line) and experimental frequency responses obtained in the absence of an external magnetic field. (Reproduced with permission from R. Ciocan and N. Ida, "Transmission line matrix model for detection of local changes in permeability using a microwave technique," in *IEEE Transactions on Magnetics*, vol. 40, no. 2, pp. 651–654, March 2004. doi: 10.1109/TMAG.2004.824883. © 2004 IEEE)

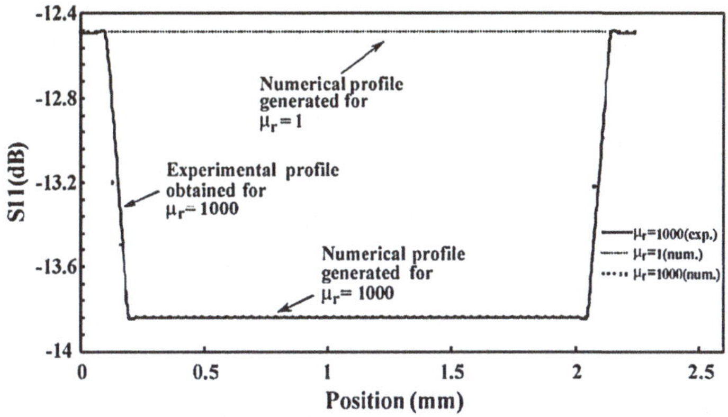

Fig. 5 Experimental and numerical results for a 2-mm sample produced by scanning across the sample. (Reproduced with permission from R. Ciocan and N. Ida, "Transmission line matrix model for detection of local changes in permeability using a microwave technique," in *IEEE Transactions on Magnetics*, vol. 40, no. 2, pp. 651–654, March 2004. doi: 10.1109/TMAG.2004.824883. © 2004 IEEE)

avoid distortions in the field. Large samples require considerable time for scanning especially if two-dimensional scans are called for, pointing to a serious limitation of the method. Nevertheless, under the right conditions this method is very accurate and the resolution extremely high.

3 Resonant Testing Methods

For production-level high sensitivity testing, the use of resonant methods is one of the best approaches, provided that the conditions of testing and the samples being tested lend themselves to this type of test. When resonant methods can be used for testing, they offer high resolution, high sensitivity, and an ease of signal analysis that cannot be matched in other methods. In most cases what is measured is the resonant frequency that is essentially noise free and extremely easy to measure by digital means. In the process of measuring the resonant frequency, scanning is done over a range of frequencies and therefore the frequency response of the resonator over the range is obtained. This then allows the calculation of the quality factor of the cavity. In a resonator, regardless of its physical construction, the resonant frequency is primarily a function of the physical dimensions and the real part of permittivity and permeability of the medium within the resonator. The quality factor is directly related to the losses in the cavity and therefore the quality factor is a measure of these losses through the imaginary parts of the complex permittivity and permeability. By measuring both the resonant frequency and the quality factor, a complete characterization of the material properties can be obtained and through that the condition of the test sample, including effects such as cracks, stresses in materials, delaminations, inclusions, and the like. In effect what is measured is the effective complex permittivity and the effective complex permeability. There are a variety of resonant structures that can be used for testing. The most obvious is the cavity resonator in which the sample is either inserted in the cavity (Fig. 6a) or the cavity is partially opened to allow samples either to pass through the cavity or the cavity to pass over the sample (Fig. 6b). The configuration in Fig. 6a is limited to small samples that can be inserted in the cavity and hence it is used mostly for evaluation of material properties under specific, controlled conditions. This may mean, for example, that the sample must have a certain size or shape or that it fills a portion of the cavity in which conditions are appropriate for the test. In general this is a limitation both in terms of the types of materials and conditions that can be evaluated and in terms of the speed and accuracy of the test. In partially open cavities, the range of materials and structures can be extended to flat samples, tubes and bars or even to continuous testing of fabrics and polymers on the production

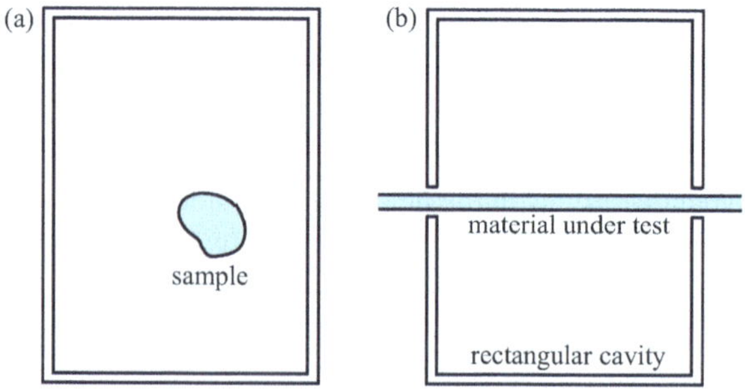

Fig. 6 Testing in resonant cavities. (**a**) A small sample introduced into the cavity. (**b**) Testing of flat sample in a partially open rectangular cavity

line. In that respect, the most commonly measured quantity is moisture content but also density, permittivity, dimensions and any condition related to these.

Resonators can also be made of transmission lines, which are often open [30, 57, Pozar 1998] and can therefore easily sense conditions in their proximity. An application of this type resonator to online industrial testing and monitoring will be given in the following section.

Equation (2) describes the shift in resonant frequency for a cavity in which the permeability and permittivity are uniform throughout the cavity and the change $\Delta\mu$ and $\Delta\varepsilon$ is also uniform (Fig. 6b). This occurs in testing of fluids or gases that fill the cavity, such as in monitoring of humidity or for detection of gaseous emissions from explosives. Although Eqs. (1) and (2) simply indicate permittivity and permeability, both can be complex to allow for modeling of complex media:

$$\varepsilon_c = \varepsilon' + j\varepsilon'' = \varepsilon' - j\frac{\sigma_s}{\omega} \tag{4}$$

Equation (1) now becomes [30]

$$\frac{f - f_0}{f_0} = -\frac{\int_v \left[\left(\Delta\varepsilon' - j\frac{\Delta\sigma}{\omega}\right)\mathbf{E}\cdot\mathbf{E}_0 - \Delta\mu\mathbf{H}\cdot\mathbf{H}_0\right]dv}{\int_v \left[\varepsilon_0\mathbf{E}\cdot\mathbf{E}_0 - \mu_0\mathbf{H}\cdot\mathbf{H}_0\right]dv} \tag{5}$$

The change in permeability may be real or complex or may be zero. ε_0 and μ_0 are the properties of the empty cavity.

A quality factor is defined as the peak stored energy divided by the energy dissipated per period as an indication of performance of the cavity:

$$Q = 2\pi\frac{\text{Peak stored energy}}{\text{Energy loss per period}} \tag{6}$$

If we assume for simplicity that power loss occurs only in the volume of the cavity (none in its walls), the Q-factor is:

$$Q = \frac{\int_v \varepsilon_0 {\mathbf{E}_0}^2 dv}{\int_v \left[\varepsilon''\mathbf{E}\cdot\mathbf{E}_0\right]dv} \tag{7}$$

This shows the physical properties that affect the Q-factor but in practice the Q-factor is calculated from the half-power points of the frequency response as:

$$Q = \frac{2\pi f_0}{BW} = \frac{f_0}{f_u - f_l} \tag{8}$$

f_u is the upper half power frequency and f_l the lower half power frequency of the resonating cavity. In a practical test, the frequency response is obtained by measuring the S_{11} parameter using a network analyzer, and by measuring the amplitude of the response, the upper and lower half-power frequencies are obtained.

The relations above apply to closed, partially open, or entirely open resonator of any kind. The methods based on perturbation are most easily applied to closed cavities whereas those based on measurement of resonant frequency and Q-factor are universal since they do not actually depend on the physical attributes of the structure—only on their frequency response.

4 Testing Through Monitoring of Moisture Content

As an example of the use of open transmission line resonators for testing of moisture content, the monitoring of moisture in production of latex-coated fabrics for the production of tires is considered. Fabrics such as nylon, polyester, or aramid are coated with latex for the production of tire belts. The amount of latex is critical. Too much or too little and the bonding process suffers. In addition, during the coating process, streaking, dry patches, and breaks in the fabric must be detected to prevent failure of tires. A resonant sensor suitable for this purpose is shown in Fig. 7. The sensor consists of two metal shields and two center plates (dimensions shown in the figure). Two short probes penetrate through the shield to couple energy to the center plates and drive the resonator. These are connected to a network analyzer, which supplies the power needed and measures the response by measuring the reflection (S_{11}) and transmission (S_{21}) parameters of the resonator to obtain the resonant frequency and the Q-factor continuously while the fabric moves at production speeds. The fabric is thin (ranging in thickness from 0.3 to 0.8 mm. Production requirements are typically for about 5% solid latex (measured after drying) while in the wet stage the latex is 20% of the solution by volume. The sensor measures the permittivity in the wet stage.

This sensor is called a broadside coupled stripline resonator [30, 58]. It is particularly useful since it resonates in two distinct modes. The even mode is sensitive to axial variations in permittivity and is used in this application to measure

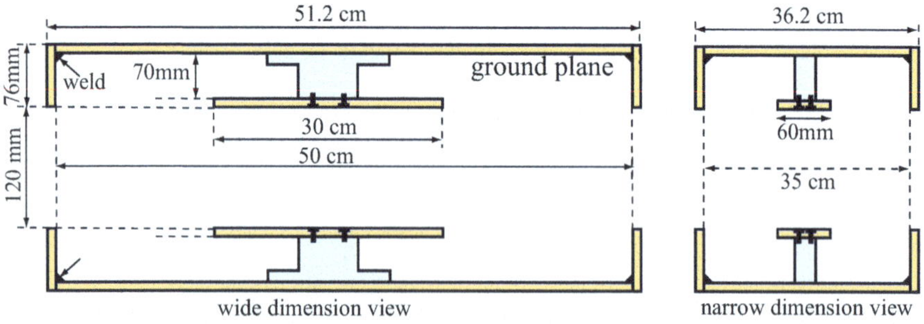

Fig. 7 Open transmission line sensor with dimensions. (Reproduced with permission from Open Resonator Microwave Sensor System for Industrial Gauging, N. Ida, The Institution of Engineering and Technology, 2018, ISBN 978-1-78561-140-7)

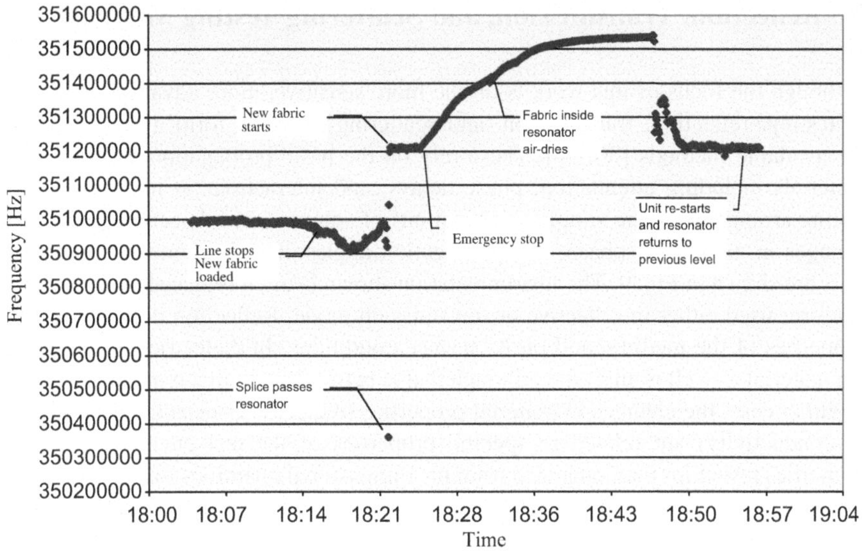

Fig. 8 Resonant frequency measured over time. (Reproduced with permission from Open Resonator Microwave Sensor System for Industrial Gauging, N. Ida, The Institution of Engineering and Technology, 2018, ISBN 978-1-78561-140-7)

moisture content. The odd mode is sensitive to variations in directions vertical to the axis of motion of the fabric and is used for compensation of variations in resonant frequency due to temperature and moisture in the sensor.

Figure 8 shows some of the capabilities of the sensor as recorded by the network analyzer on the production line. On the left, the production line is stopped to introduce a different fabric. The lowering of the resonant frequency indicates that during the changeover process the two fabrics, which are attached to each other, present a higher resonant frequency. As the line starts after the changeover, the passing of the stitch is seen as a dot at a much lower resonant frequency (the stitch is an overlap of the two fabrics). Then the new fabric, that either has a lower permittivity or absorbs less of the latex solution shows a higher frequency than the first fabric. Some time after, the line stops and the fabric dries in air. While stopped, the resonant frequency increases (lower permittivity). As soon as the line starts, the resonant frequency returns to the previous value indicating the permittivity of the fabric.

A sensor is only as good as its sensitivity and resolution. The sensor described here has a resolution below 0.02% (for dry latex) given a 10 kHz resolution of the network analyzer. This can be increased at lest by a factor of 10 since the network analyzer is capable of resolving frequencies below 100 Hz but the figure of approximately 0.1% is more than sufficient in this application.

5 Reflection, Transmission, and Scattering Testing Methods

Although the focus of this work is on the more sensitive, more advanced methods of testing, reflection, transmission, and scattering methods form a large part of the available methods [30, 55]. These rely on the basic propagation properties of materials including attenuation, phase delays, and interactions at interfaces. The testing arrangements are simple combinations of antennas and detectors, recording changes in amplitude, phase, and propagation angles. The fundamentals of these tests are shown in Fig. 9. The instrumentation shown is often replaced by a network analyzer used either in reflective or transmission mode. Reflection depends on the properties of the medium and on its surface conditions. In dielectrics, the bulk of the material as well is analyzed although that is better done with a transmission test. In either case, the changes in material properties (dielectric constant, permeability, or conductivity) are related to specific properties of the test environment. The properties tested for may range from simple dimensional quantities to delamination, humidity, contamination, and a host of other conditions. In practice, almost any material property may be related to a change in its electrical properties and hence detectable through transmission or reflection tests. In some tests, a single antenna is used in pulsed mode and serves both for transmission and for reception of the signal in a method akin to pulsed radar although reflection tests are often done at short range. Reflection and transmission can be combined by using a reflector on the opposite side of the sample and recording both the directly reflected signal and the signal that passes twice through the sample. This is particularly useful in cases the reflection coefficient is small.

An example of the use of reflections from an inclusion in a lossy dielectric medium is shown in Fig. 10. This is yet another variation on reflection methods in that use is made of two receivers and the signal of one is subtracted from the other

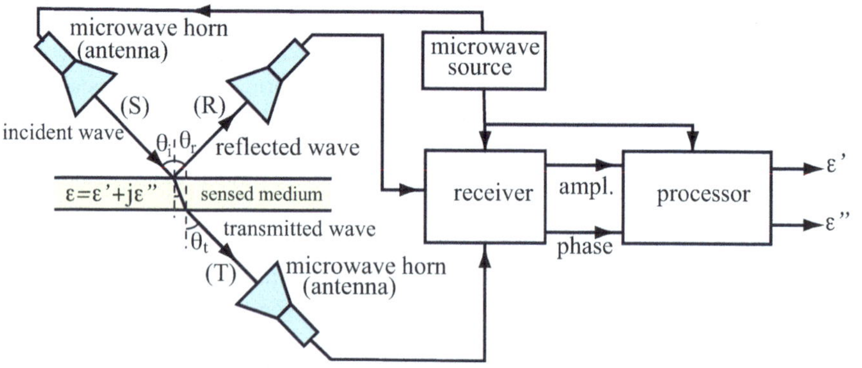

Fig. 9 Test method for the complex permittivity of a medium. Either the reflection or transmission method may be used

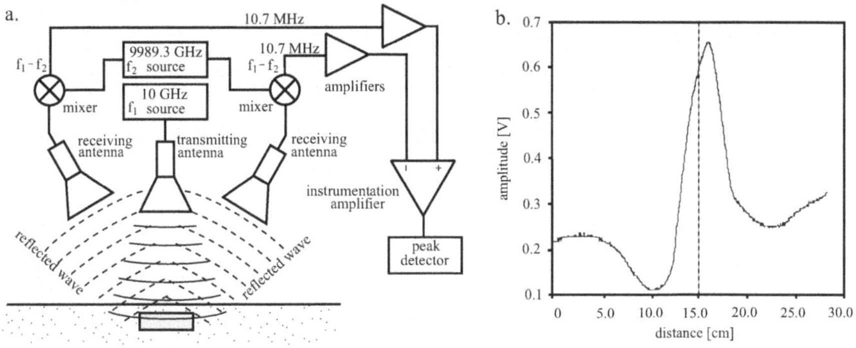

Fig. 10 (**a**) Detection of buried plastic (dielectric) mines using a differential reflection method. (**b**) The signal obtained from a scan across the dielectric shows a peak at the center of the dielectric and two dips indicating its corners

to obtain a differential sesor. This results in reduced noise levels and in emphasis on changes in permittivity. It therefore is more sensitive to the edges of the inclusion.

The configuration in Fig. 10 has been used as a simple, near-surface ground penetrating radar (GPR) but the basic idea also has applications in mine detection or buried pipelines and other utility structures or in archaeology. The transmitted (center) signal is low (about 100 mW) at approximately 10 GHz. The transmitter and receiver are at the same height from the surface and can vary according to needs. The signal in Fig. 10b was obtained from a height of 30 cm but higher resolution may be obtained at closer range. The transmitted signal penetrates into the groung and is reflected off any discontinuity such as buried objects. A large reflection due to the surface of the ground is also present. The two receivers receive the signals and these are subtracted to eliminate common mode features such as the reflection due to the ground surface. For this reason, the edges of the object in Fig. 10a can be seen as large dips in Fig. 10b whereas away from the object the signal is low. In this particular case, the wavelength is approximately 3 cm hence the expected resolution is of this order.

6 Radar Methods

Radar and radar-like methods in NDT can be traced to the first use of scattered fields for testing. An early use can be seen in Fig. 11. In an attempt to increase sensitivity to small defects and delaminations in composites and dielectrics, which provide little changes in either reflection or transmission, the scattered field at right angle is used for detection. Although this field is also small it only exists if a flaw exists and hence its detection is improved. In Fig. 11, antenna B can be used for a transmission test whereas antenna C is used for scattering text. If the signal is pulsed, antenna A can also be used for reflection testing.

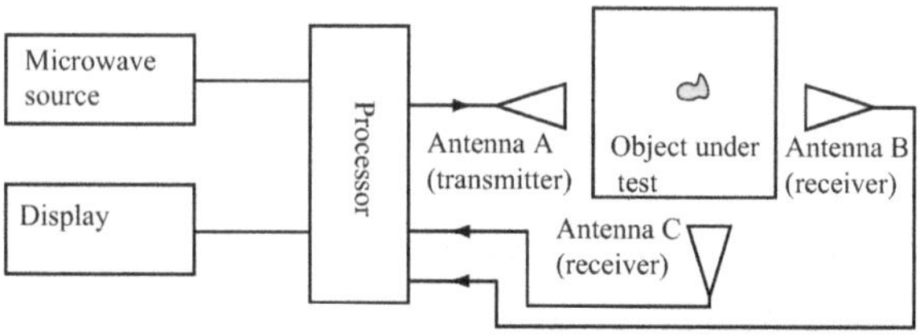

Fig. 11 Detection of defects or inclusions by a scattering method

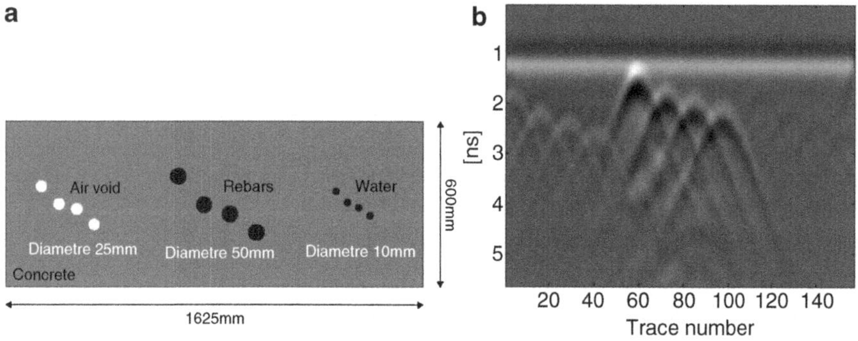

Fig. 12 Simulation of GPR scanning of concrete. (**a**) Various conditions at varying depts. From the surface. (**b**) Signal map of the interaction of waves in the concrete. The thick dark bar is the surface of the concrete [14]

True radar testing has been applied to evaluation of concrete using GPR equipment and techniques. The purpose is detection and evaluation of conditions such as inclusions and cracking but primarily evaluation of the condition of reinforcement steel. This is an important aspect of testing since deterioration of reinforcement bars due to corrosion leads to weakening of the concrete structure and even failure. A simulation of conditions in concrete and the results that can be obtained using a GPR scan can be seen in Fig. 12. It is clear that rebars, being made of steel, will result in large reflections but water inclusions or air pockets are equally detectable in spite of the relatively high loss of concrete. The advantage of using GPR for this type of testing is the speed with which large areas of concrete can be covered. Considering the amount of concrete in infrastructures, this is an overriding condition.

A promising newer approach to radar for use in NDT is the ultra wide band (UWB). UWB refers to the use of a wide bandwidth with reduced power and power density. Typically, a very narrow pulse is transmitted generating the wide spectrum typical of the method. The bandwidth is normally about 25% of the center frequency. UWB radars are very compact, transmit an effective radiated isotropic power (ERIP) of less than –41.3 dB/MHz in the range between 3.1 and 10.6 GHz. The wide bandwidth allows interaction with flaws and material properties over a wide range of frequency, akin to multifrequency testing. Because of the low power and the spread in spectrum, regulatory constraints are minimal, interference is almost nonexistent and resolution increases. The main problem with this type of radar is worsening of the signal–to-noise ratio. From the operational point of view, these radars are inexpensive, available as single-board systems, and some of the signal processing algorithms are also available. Although UWB radars are not new, they have only become small enough and practical in the last few years.

The uses of these radars are still evolving. A common application is colloquially called "see through wall" [16, 59, Pochanin et al. 2024]. Just as in GPR, the idea is to obtain and quantify signals from objects and conditions located behind or within lossy dielectrics. Surface defects have similarly been analyzed using comparative signals [16]. The method has also been applied to aeronautic component testing [60] as well as detection of rebar in concrete (Xu et al. 2013). Other applications include measurement of conditions of walls in salt mines [61] in which depths up to 2 m could be analyzed, as well as others [16, Zhang et al. 2014]. UWB radars operate at low powers and offer high resolution consistent with the high-frequency content of the signals. As such, forms of UWB radars have been applied to medical diagnostics, with particular preference to breast cancer [24–27]. In this particular application, the UWB radar takes advantage of the relatively high contrast between healthy and malignant tissue.

In NDE proper, UWB radar is still in the experimental stages. These attempts use mostly experimental devices [20, 23, Shi et al. 2016, Kidera et al. 2014, Yang et al. 2008].

7 Other Methods

Microwave application in NDT are not limited to those described above. Another method that is used, to a much smaller extent, is microwave radiometry whereby either the background microwave radiation is used (passive radiography) or a source is available (active radiography) [62, 63]. In either case, a radiometer (a device that measure the radiant flux of the source after it passes through the sample) is used. Here too, the types of sensors that can be employed vary and may include bolometers, diodes, or calorimeters to measure the radiant flux [64]. In radiometric microwave methods, the radiated power and its distribution is measured after the radiant flux passes through the test sample. This power is measured by common methods such as bolometers (or bolometer arrays) or diodes. In some cases, calorimetric methods are used. In bolometers and calorimeters, the rise in temperature is measured whereas diodes measure power (or power density) directly.

Although active or passive radiometry may be used, the preference in NDE is for active methods in which the sources can be accurately controlled. In applications where imaging is preferred, either an array of sensors or a scanning sensor is used [65–67]. For surface effects such as surface roughness or flaws, including surface-breaking and near subsurface effect, reflective methods can also be used [68–71]. In general, the higher-frequency ranges are preferred for improved resolution but this is somewhat difficult to attain at present because of the limited availability of sources above about 100 GHz [62, 72, Bakhtiari et al. 2012].

Microwaves are also used as sources of power for other methods of NDT. One common example of this type of use is for pulsed thermography in which the heat is provided by microwave pulses.

References

1. Liskow, C. L. (1948). *Ultra-high frequency apparatus for inspection of sheet and other materials*, US patent 2,457-895.
2. Bakhtiari, S., Qaddoumi, N., Ganchev, S. I., & Zoughi, R. (1994). Microwave noncontact examination of disbond and thickness variation in stratified composite media. *IEEE Transactions on Microwave Theory and Techniques, 42*, 389–395.
3. Decreton, M., & Gardiol, F. E. (1974). Simple nondestructive method for measurement of complex permittivity. *IEEE Transactions on Instrumentation and Measurement, 23*, 434–438.
4. Ghodgaonkar, D. K., & Ali, N. A. (2000). Microwave nondestructive testing of composite materials using free-space microwave measurement techniques (pp. 15–21). In *Proceedings of 15th World Conference on Non-Destructive Testing*.
5. Bahr, A. J. (1978). Microwave NDE of ceramics (pp. 236–241). In *Proceedings of the ARPA/AFML Review of Progress in Quantitative NDE*.
6. Hughes, D., & Zoughi, R. (2005). A novel method for determination of dielectric properties of materials using a combined embedded modulated scattering and near-field microwave techniques. Part I—Forward model. *IEEE Transactions on Instrumentation and Measurement, 54*, 2389–2397.
7. Kharkovsky, S., Akay, M. F., Hasar, U. C., & Atis, C. D. (2002). Measurement and monitoring of microwave reflection and transmission properties of cement-based specimens. *IEEE Transactions on Instrumentation and Measurement, 51*, 1210–1218.
8. Mukherjee, S., Tamburrino, A., Haq, M., Udpa, S., & Udpa, L. (2018). Far field microwave NDE of composite structures using time reversal mirror. *NDT and E International, 93*, 7–17.
9. Stuchly, M., & Stuchly, S. (1980). Coaxial line reflection methods for measuring dielectric properties of biological substances at radio and microwave frequencies—A review. *IEEE Transactions on Instrumentation and Measurement, 29*, 176–183.
10. Bogosanovic, M., Al Anbuky, A., & Emms, G. W. (2012). Microwave nondestructive testing of wood anisotropy and scatter. *IEEE Sensors Journal, 13*, 306–313.
11. Caorsi, S., Massa, A., Pastorino, M., & Randazzo, A. (2003). Electromagnetic detection of dielectric scatterers using phaseless synthetic and real data and the memetic algorithm. *IEEE Transactions on Geoscience and Remote Sensing, 41*, 2745–2753.
12. Han, J., & Nguyen, C. (2007). Development of a tunable multiband UWB radar sensor and its applications to subsurface sensing. *IEEE Sensors Journal, 7*, 51–58.
13. Joongsuk, & Nguyen, C. (2005). An ultrawide-band microwave radar sensor for nondestructive evaluation of pavement subsurface. *IEEE Sensors Journal, 5*, 942–949.
14. Travassos, L. (2007). *Modélisation numérique pour l'évaluation non destructive électromagnétique : application au contrôle non destructif des structures en béton*, PhD dissertation, École Centrale de Lyon, France.
15. Yang, X., Zheng, Y. R., Ghasr, M. T., & Donnell, K. M. (2017). Microwave imaging from sparse measurements for near-field synthetic aperture radar. *IEEE Transactions on Instrumentation and Measurement, 66*, 2680–2692.
16. Gubinelli, S., Paolini, M., Andrea Giorgetti, A., Troiani, E., & Chiani, M. (2014). An ultra-wideband radar approach to nondestructive testing. *IEEE International Conference on Ultra-WideBand (ICUWB)*. https://doi.org/10.1109/ICUWB.2014.6958997.
17. Zhong, C. H., Croxford, A. J., & Wilcox, P. D. (2014). Remote inspection system for impact damage in large composite structure. *Proceedings of the Royal Society A: Mathematical, Physical and Engineering Sciences, 471*, 20140631.

18. Xie, Y., Guo, B., Xu, L., Li, J., & Stoica, P. (2006). Multistatic adaptive microwave imaging for early breast cancer detection. *IEEE Transactions on Biomedical Engineering, 53*, 1647–1657.
19. Xu, L., Davis, S., Hagness, S. C., Weide, D., & Veen, B. (2004). Microwave imaging via space–time beamforming: Experimental investigation of tumor detection in multilayer breast phantoms. *IEEE Transactions on Microwave Theory and Techniques, 52*, 1856–1865.
20. Pochanin, G., Masalov, S., Pochanina, I., Capineri, L., Falorni, P., & Bechtel, T. (2016). *Modern Trends in Development and Application of the UWB Radar Systems 8th International Conference on Ultrawideband and Ultrashort Impulse Signals (UWBUSIS)*. https://doi.org/10.1109/UWBUSIS.2016.7724141.
21. Shin, H. J., Narayanan, R. M., Asmuth, M. A., & Rangaswamy, M. (2016). Ultrawideband noise radar tomography: Principles, simulation, and experimental validation. *International Journal of Microwave Science and Technology, 2016*. https://doi.org/10.1155/2016/5787895.
22. Kidera, S., Sakamoto, T., & Sato, T. (2010). Accurate UWB radar three-dimensional imaging algorithm for a complex boundary without range point connections. *IEEE Transactions on Geoscience and Remote Sensing, 48*, 1993–2004.
23. Li, Z., & Meng, Z. (2016). A review of the radio frequency non-destructive testing for carbon-fibre composites. *Measurement Science Review, 16*, 68–76. https://doi.org/10.1515/msr-2016-0010.
24. Bidhendi, H. K., Jafari, M. H., & Genov, R. (2014). Ultra-wideband imaging systems for breast cancer detection. In M. Yuce (Ed.), *Ultra-wideband and 60 GHz communications for biomedical applications*. NewYork: Springer. https://doi.org/10.1007/978-1-4614-8896-5_5.
25. Fear, E. C., Hagness, S. C., Meaney, P. M., Okoniewski, M., & Stuchly, M. A. (2002a). Enhancing breast tumor detection with near-field imaging. *IEEE Microwave Magazine, 3*, 48–56.
26. Fear, E. C., Li, X., Hagness, S. C., & Stuchly, M. A. (2002b). Confocal microwave imaging for breast cancer detection: Localization of tumors in three dimensions. *IEEE Transactions on Biomedical Engineering, 49*, 48–56.
27. Joines, W. T., Zhang, Y., Li, C., & Jirtle, R. L. (1994). The measured electrical properties of normal and malignant human tissues from 50 to 900 MHz. *Medical Physics, 21*, 547–550.
28. Auld, B. A. (1978). *Theory of ferromagnetic resonance probes for surface cracks in metals*. Stanford: E.L. Ginzton Lab., Stanford University. G.L. rep. 2839.
29. Ida, N. (2008). Open stripline resonator sensors for rubber properties gauging (pp. 1077–1083). In *Proceedings of MOMAG-08*, Florianopolis, Brazil, September 7–10, 2008.
30. Ida, N. (2018). *Open resonator microwave sensor systems for industrial gauging: A practical design approach*. London: IET.
31. Ida, N., & Bhuyia, O. (2008). Design and optimization of an open stripline resonator for rubber thickness gauging (Vol. 4, pp. 97–100). In *Proceedings of the 11th International Conference on Optimization of Electrical and Electronic Equipment*, Brasov, Romania.
32. Joffe, R., Shavit, R., & Kamenetskii, E. (2017). Multiresonance measurement method for microwave microscopy. *IEEE Transactions on Instrumentation and Measurement, 66*, 2174–2180.
33. Li, Y., Bowler, N., & Johnson, D. B. (2011). A resonant microwave patch sensor for detection of layer thickness or permittivity variations in multilayered dielectric structures. *IEEE Sensors Journal, 11*, 5–15.
34. Ciocan, R., & Ida, N. (2004). Transmission line matrix model for detection of local changes in permeability using a microwave technique. *IEEE Transactions on Magnetics, 40*, 651–654.
35. Qaddoumi, N. N., Saleh, W. M., & Abou-Khousa, M. (2007). Innovative near-field microwave nondestructive testing of corroded metallic structures utilizing open-ended rectangular waveguide probes. *IEEE Transactions on Instrumentation and Measurement, 56*, 1961–1966.
36. Anlage, S. M., Talanov, V. V., & Schwartz, A. R. (2007). Principles of Near-field microwave microscopy. In S. V. Kalinin & A. Gruverman (Eds.), *Scanning probe microscopy: Electrical and electromechanical phenomena at the nanoscale* (Vol. 1, pp. 215–253). New York: Springer.

37. Ciocan, R. (2000). *Evanescent microwave microscope: a novel nondestructive, high resolution method for quantitative evaluation of materials*. MS thesis, Case Western Reserve University, pp. 39–46.
38. Ghasr, M. T. (2004). *Detection and size evaluation of corrosion precursor pitting using near-field microwave and millimeter wave nondestructive testing methods*. Master's thesis, Electrical and Computer Engineering Department, University of Missouri-Rolla, Rolla, MO.
39. Akuthota, B., Hughes, D., Zoughi, R., & Myers, J. (2004). Near-field microwave detection of disbond in fiber reinforced polymer composites used for strengthening concrete structures and disbond repair verification. *ASCE Journal of Materials in Civil Engineering, 16*, 540–546.
40. Ash, E. A., & Nicholls, G. (1972). Super-resolution aperture scanning microscope. *Nature, 237*, 510–512.
41. Chen, G., Hu, B., Takeuchi, I., Chang, K. S., Xiang, X. D., & Wang, G. (2005). Quantitative scanning evanescent microwave microscopy and its applications in characterization of functional materials libraries. *Measurement Science and Technology, 16*, 248–260. https://doi.org/10.1088/0957-0233/16/1/033.
42. Bakhtiari, S., & Zoughi, R. (1990). Microwave thickness measurement of lossy layered dielectric slabs using incoherent reflectivity. *Research in Nondestructive Evaluation, 2*, 157–168.
43. Ganchev, S., Qaddoumi, N., Brandenburg, D., Bakhtiari, S., Zoughi, R., & Bhattacharyya, J. (1994). Microwave diagnosis of rubber compounds. *IEEE Transactions on Microwave Theory and Techniques, 42*, 18–24.
44. Bois, K. J., Benally, A., & Zoughi, R. (2000). Microwave near-field reflection property analysis of concrete for material content determination. *IEEE Transactions on Instrumentation and Measurement, 49*, 49–55.
45. Peer, S., Case, J. T., Gallaher, E., Kurtis, K. E., & Zoughi, R. (2003). Microwave reflection and dielectric properties of mortar subjected to compression force and cyclically exposed to water and sodium chloride solution. *IEEE Transactions on Instrumentation and Measurement, 52*, 111–118.
46. Sayar, M., & Ogawa, D. S. K. (2009). Non destructive microwave detection of layer thickness in degraded thermal barrier coatings using K-and W-band frequency range. *NDT&E International, 42*, 398–403.
47. Anderson, P. L. (1997). Development of coating thickness instrument for graphite/epoxy composites. In *Review of progress in quantitative nondestructive evaluation* (Vol. 16, pp. 1601–1607). New York: Plenum.
48. Ghasr, M. T., Horst, M. J., Lechuga, M., Rapoza, R., Renoud, C. J., & Zoughi, R. (2015). Accurate one-sided microwave thickness evaluation of lined-fiberglass composites. *IEEE Transactions on Instrumentation and Measurement, 64*, 2802–2812.
49. Zoughi, R., Gallion, J. R., & Ghasr, M. T. (2016). Accurate microwave measurement of coating thickness on carbon composite substrates. *IEEE Transactions on Instrumentation and Measurement, 65*, 951–953.
50. Mazlumi, F., Sadeghi, S. H. H., & Moini, R. (2006). Interaction of an open-ended rectangular waveguide probe with an arbitrary shape surface crack in a lossy conductor. *IEEE Transactions on Microwave Theory and Techniques, 54*, 3706–3711.
51. Rosner, B. T., & Van der Weide, D. W. (2002). High-frequency near field microscopy. *Review of Scientific Instruments, 73*, 2505–2525.
52. Acher, O., Baclet, P., & Perin, G. (1996). Investigation of the microwave permeability of ferromagnetic film laminations under a magnetic field. *Journal of Magnetism and Magnetic Materials, 157/158*, 480–481.
53. Ustinov, V. V., Rinkevich, A. B., & Romashev, L. N. (1999). Microwave magnetoresistance of Fe/Cr multilayers in current-perpendicular to plane geometry. *Journal of Magnetism and Magnetic Materials, 198/199*, 82–84.
54. Tabib-Azar, M., Garcia-Valenzuela, A., & Ponchak, G. (2002). *Evanescent microwave microscopy for high resolution characterization of materials*. Norwell, MA: Kluwer.
55. Ida, N. (1992). *Microwave NDT* (pp. 201–273). Dordrecht: Kluwer Press.

56. Ciocan, R., Ida, N., Ciocan, E., & Jiang, H. (2004). Applications of the transmission line matrix method to microwave scanning microscopy. *ACES Journal, 19*, 94–100.
57. Ida, N. (2015). *Engineering electromagnetics*. Switzerland: Springer.
58. Garg, R., Bahl, I., & Bozzi, M. (2013). *Microstrip lines and slotlines*. Boston: Artech House.
59. Yang, Y. (2008). *Development of a real-time ultra-wideband see through wall imaging radar system*. PhD dissertation, University of Tennessee, Knoxville TN.
60. Cristofani, E., Friederich, F., Vandewal, M., & Jonuscheit, J. (2017). Nondestructive testing of aeronautics composite structures using ultrawideband radars. In J. D. Taylor (Ed.), *Signals, targets, and applications* (pp. 237–256). Boca Raton: CRC Press.
61. Herrmann, R. (2011). *M-sequence based ultra-wideband radar and its application to crack detection in salt mines*, PhD dissertation, Ilmenau University of Technology, Department of Electrical Engineering and Information Technology. URN: urn:nbn:de:gbv:ilm1-2011000344.
62. Appleby, R., Anderton, R. N., Thomson, N. H., & Jack, J. W. (2004). The design of a real-time 94-GHz passive millimetre-wave imager for helicopter operations. *Passive Millimetre-Wave and Terahertz Imaging and Technology, 5619*, 38–46. https://doi.org/10.1117/12.581336.
63. Ulaby, F. T., Moore, R. K., & Fung, A. F. (1981). *Microwave remote sensing - Active and passive - Microwave Remote sensing fundamentals and radiometry* (Vol. 1, pp. 344–418). Reading, MA: Addison-Wesley.
64. McCluney, W. R. (2014). *Introduction to radiometry and photometry* (2nd ed.). Boston: Artech House.
65. Bakhtiari, K., Gopalsami, N., & Raptis, C. (1997). Determining bonding quality in polymer composites with a millimeter wave sensor. In *Review of progress in quantitative nondestructive evaluation* (pp. 665–671). New York: Springer.
66. Shibuya, T., Goto, Y., Dobroiu, A., Otani, C., & Kawase. (2007). Nondestructive detection of carbon in soot collection filters by using a 94 GHz source (pp. 273–274). In *Joint 32nd International Conference on Infrared and Millimeter Waves and the 15th International Conference on Terahertz Electronics*, Cardiff, UK
67. Viegas, C., Alderman, B., Huggard, P. G., Powell, J., Parow-Souchon, K., Firdaus, M., Liu, H., Duff, C. I., & Sloan, R. (2017). Active millimeter-wave radiometry for nondestructive testing/evaluation of composites—Glass fiber reinforced polymer. *IEEE Transactions on Microwave Theory and Techniques, 65*, 641–650.
68. Clancy, T., Ulaby, F., & Dobson, C. (2000). Detection of Space Shuttle tile defects using millimeter-wave radiometry (Vol. 3, pp. 1787–1790). In *IEEE Antennas and Propagation Society International Symposium*, Salt Lake City, UT.
69. Harmer, S. W., Shylo, S., Shah, M., Bowring, N. J., & Owda, A. Y. (2016). On the feasibility of assessing burn wound healing without removal of dressings using radiometric millimetre-wave sensing. *Progress in Electromagnetics Research M, 45*, 173–183.
70. Seah, K. H., Karim, M. F., Ong, L. C., & Chiam, T. M. (2012). Rapid corrosion detection using 94 GHz millimeter wave technology. In *IEEE International Instrumentation and Measurement Technology Conference (I2MTC)* (pp. 473–476). Austria: Graz.
71. Smulders, P. F. M. (2012). Analysis of human skin tissue by millimeter-wave reflectometry. *Skin Research and Technology, 19*, 209–216. https://doi.org/10.1111/j.1600-0846.2012.00629.x.
72. Bardati, F., Brown, V. J., Ross, M. P., & Tognolatti, P. (1992). Microwave radiometry for medical thermal imaging: Theory and experiment. In *IEEE MTT-S International in Microwave Symposium Digest* (Vol. 3, pp. 1287–1290). https://doi.org/10.1109/MWSYM.

A Discussion on the Algorithm Design of Electrical Impedance Tomography for Biomedical Applications

Mingyong Zhou and Hongyu Zhu

Abstract In this chapter, we present a discussion on the algorithm design of Electrical Impedance Tomography (EIT) for biomedical applications. Based on the Maxwell differential equations and the derived finite element (FE) linear equations, we first investigate the possibility to estimate the matrix that contains the impedance values based on Singular Value Decomposition (SVD) approximations. Secondly based on the biomedical properties we further explore the possibility to recover the impedance values uniquely by injecting various different types of currents with multi-frequency. Injecting various types of multi-frequency currents lead to a set of different measured voltages configurations, thus enhancing the possibility of uniquely recovering the impedance values in a stable way under the assumption that the biological cells respond to the different types of injecting currents in a different way.

By converting the Maxwell differential equations into linear equations by Finite Element (FE) method, we are able to focus on the discussions based on the linear algebra method. We also explore some insights into the biological cells' electrical properties so that we can make use of the biological cell's electrical properties to make the numerical algorithm design more stable and robust.

Keywords Electrical impedance tomography (EIT) · Finite element (FE) · Singular value decomposition (SVD) · Maxwell differential equations

1 Introduction

Electrical Impedance Tomography (EIT) draws the attention of researchers in biomedical engineering and applied mathematics for its low cost and convenient approaches using the well-known electrical method [1–4]. From the hardware

---

M. Zhou (✉) · H. Zhu
School of Computer Science and Communication Engineering, Guangxi University of Science and Technology (GXUST), Liuzhou, China

E. T. Quinto et al. (eds.), *The Proceedings of the International Conference on Sensing and Imaging, 2018*, Lecture Notes in Electrical Engineering 606,
https://doi.org/10.1007/978-3-030-30825-4_14

design including the injection of various types of currents to the imaging algorithms development invoking mathematical methods, fruitful research results of EIT appear in recent years in applied mathematics, industrial areas, and biomedical engineering. In this chapter, however we focus our discussions on the area of biomedical imaging applications. Imaging in medical application is more complex and challenging because of the complex tissues and cells structures in human bodies.

In this chapter, we focus our discussions on the algorithm design as well as the implications of the electrical properties of tissues and cells on the algorithm design. First, we outline the EIT problem as Maxwell differential equations and address the forward problem and inverse problem (imaging problem). Secondly, based on the Maxwell differential equations and the derived finite element (FE) linear equations, we investigate the possibility to estimate the matrix that contains the impedance values based on Singular Value Decomposition (SVD) approximations. It is shown that based on the SVD method the solution to the EIT problem is not unique without a prior information on the impedance values. Last but not the least, based on the biomedical properties we further explore the possibility to recover the impedance values uniquely by injecting various different types of currents with multi-frequency. Injecting various types of multi-frequency currents lead to a set of different measured voltages configurations, thus enhancing the possibility of uniquely recovering the impedance values in a numerically stable way under the assumption that the biological cells respond to the different types of injecting currents in a different way. It is worth noting that we convert the whole non-linear EIT problem into a linear problem with a very larger and very high dimension based on Finite Element (FE) method, thus allowing one to approach the EIT problem using linear algebra method especially for the biomedical applications.

We devote to our discussions on the implications of the electrical properties on the algorithm design in the last part of this chapter but not the least as is declared because this part is particularly important in the applications of biomedical engineering by taking into account the electrical properties of tissues and cells, that is, tissues and cells of human bodies respond to the different frequencies of injecting currents in a different way. We can make use of such electrical properties to ensure that the algorithm design is more stable and robust but unique simultaneously.

2 Maxwell Differential Equations and EIT Problem

Inside of magnetic field, magnetic field effect can be ignored due to the quite low permeability in biological tissue. According to the Maxwell theory and Ohm's law, we can get the description like this:

$$-\rho^{-1}\nabla\phi = \overrightarrow{J} \tag{1}$$

In this equation, ρ represents the distribution function of impedance, ϕ means potential distribution, J is the function of boundary electric current density.

Due to the fact that there is no electrical sources inside interior of biological tissue, so we get this:

$$\nabla \cdot \vec{J} = 0 \tag{2}$$

Combined Eqs. (1) and (2), we can get equation:

$$\nabla \cdot \rho^{-1} \nabla \phi = 0 \tag{3}$$

This partial differential equation should satisfy the Dirichlet boundary condition

$$\phi|_{\partial \Omega} = V \tag{4}$$

In the above equation, Ω denotes an object's boundary area. For the electrodes with inject current, Neumann boundary condition is:

$$\rho^{-1} \left. \frac{\partial \phi}{\partial \vec{n}} \right| = \vec{J} \tag{5}$$

The equations above constitute the mathematical model of EIT problem [4].

Given the impedance distributions and the injecting currents J, it is required to calculate the voltages on the surface. This is termed as a forward problem. For forward problem of EIT, we now summarize the procedure of finite element (FE) method to formulate a set of linear equations for the Maxwell differential equations represented by Eqs. (1)–(5).

3 Finite Element (FE) Method to Formulate the EIT Forward Problem

For a 3D bounded by Ω, we can divide the 3D bounded space Ω with the following triangle with K, M, N, L and coordinates represented by (x_k, y_k, z_k), $(x_M, y_{M,} z_M)$, (x_N, y_N, z_N), (x_L, y_L, z_L).

The derivation of the Finite Element (FE) method is too tedious but is a well-known and well-used method in mechanics engineering (Fig. 1). By dividing the bounded space Ω with small triangles and enough numbers of triangles, for the ith injection of currents we can formulate a linear equation as follows:

$$S_i \Phi_i = F_i \tag{6}$$

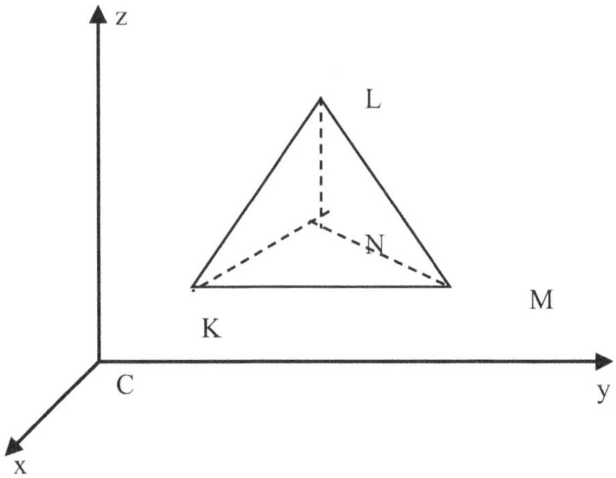

Fig. 1 Finite element triangle

where F_i is a vector that is unrelated to impedance values ρ and is related only to injecting currents, Φ_i is a vector formulated as the voltages measured on each element, and Si is a matrix formulated by the impedance values ρ of each triangle element. Now the EIT inverse problem now is to estimate the values of matrix Si, given the measured voltages vector Φ_i and vector F_i that is formulated as the injecting currents. In the following section, we present a subspace fitting approach based on SVD method to estimate the matrix Si and discuss the uniqueness of the estimation. We particularly take a Higher Order Statistics (HOS) approach to remove the Gaussian noise.

Regarding the Eq. (6), in the following section we provide a subspace fitting approach based on SVD method that is originally proposed by Petre Stoica for signal processing [5].

4 SVD Method to Solve the EIT Inverse Problem

We now re-formulate the Eq. (1) in another formation as follows by taking into the account of measurement noise denoted by $n(t)$.

$$\vec{y}_{M\times 1}(t) = A_{M\times d} \cdot \vec{x}(t) + \vec{n}(t) \tag{7}$$

The following assumptions are made:

1. Matrix A's dimension is $M \times d$, where $d < M$
2. The columns of matrix A are linearly independent
3. $\vec{n}(t)$ is a white or colored Gaussian noise by EIT measurements, and $E\left\{\vec{y}\,\vec{y}^T y_i\right\}$ as the calculated cumulant matrix from the calculations of higher

order statistics of measured data exist for all i's, where $i = 1, 2, \ldots, M$. The problem is to find all solutions of matrix A that minimizes the result of L in the following Eq. (8).

$$L = \left\| \overrightarrow{y}_{M\times 1}(t) - A \cdot \overrightarrow{x}(t) \right\|_2 \tag{8}$$

We summarize the algorithm based on SVD here as follows:

Denote the standard SVD of any cumulant matrix $E\left\{\overrightarrow{y}\,\overrightarrow{y^T}\,y_i\right\}$ as follows:

$$E\left\{\overrightarrow{y}\,\overrightarrow{y^T}\,y_i\right\} = U \sum S_{d\times M}^T \tag{9}$$

By Stoica in [5], one obtains $E\left\{\overrightarrow{y}\,\overrightarrow{y^T}\,y_i\right\}$ as follows:

$$A D_i A^T G = U \sum S^T G = 0 \tag{10}$$

where $D_i = \sum_{j=1}^{d} A_{ij} E\left\{\overrightarrow{x}\,\overrightarrow{x^T}\,x_j\right\}$, $(i = 1, 2, \ldots, M)$. Under the assumptions of (A), (B), and (C), the approximation problem becomes

$$\min_A \min_C \| A - SC \|_F \tag{11}$$

In reference [5], the optimal solutions are derived as the d^2 eigenvectors that correspond to the least minimum d^2 eigenvalues of the following matrix Q

$$Q \triangleq \left\{ I - \left(\underset{M\times d}{S} \otimes \underset{d\times d}{I} \right) \left[(S\otimes I)^u \, (S\otimes I) \right]^{-1} (S\otimes I)^H \right\} \tag{12}$$

where $\otimes$ denotes Kronecker product.

5 A Simple Experiment to Demonstrate SVD Method

We carried a simple example experiment to demonstrate the method based on SVD method. First the algorithm is outlined as follows. Note that second order statistics are used for the simple experiment and measurement noise is not taken into account in the case.

1.

$$\overrightarrow{y}_{M\times 1}(t) = A_{M\times d} \cdot \overrightarrow{x}_{d\times 1}(t) \quad (M \geq d) \tag{13}$$

where $\overrightarrow{y}(t)$ is a vector formulated by the measured voltages.

2.

$$E\left\{\vec{y}(t)\cdot\vec{y}^{T}(t)\right\}\triangleq Y_{M\times M} \tag{14}$$

where $Y_{M \times M}$ is a correlation matrix that is formulated based on second order statistics.

3. Denote SVD of cumulant matrix by:

$$Y_{M\times M}=U_{M\times d}\Sigma_{d\times d}S_{d\times M}^{T} \tag{15}$$

4. Calculate the matrix Q as follows:

$$Q\triangleq\left\{I-\left(\underset{M\times d}{S}\otimes\underset{d\times d}{I}\right)\left[(S\otimes I)^{T}(S\otimes I)\right]^{-1}(S\otimes I)^{T}\right\} \tag{16}$$

The eigenvectors of Q with dimension of $Md \times 1$ thus corresponding to the least minimum eigenvalues of Q form all possible solutions to vector $\vec{A}_{Md\times 1}$, and $\vec{A}_{Md\times 1}$ is a column vector with dimension $Md \times 1$ derived from matrix A.

One should note that the solutions to matrix A are not unique by SVD method as proposed as the above.

A simple experiment was carried out as demonstrated in Fig. 2, where the circles are placed with resistors *A*-*E* and only one resistor is placed in the center of circle.

The four voltages at point B, C, D, E with reference to point A are measured as a 4×1 data vector, that is $M = 4$. Algorithm's steps (1–5) are used for the estimation in which $d = 3$ is used. The derived Q in this experiment is thus a matrix with rows of 12 and columns of 9 (12×9).

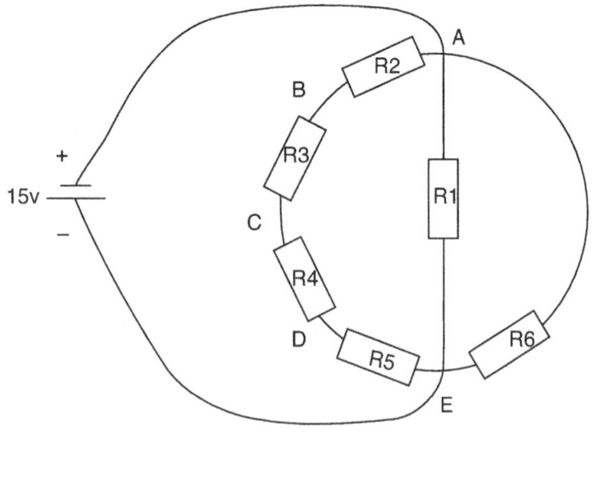

Fig. 2 An EIT experiment based on SVD method

The estimation solutions to matrix A are derived as the following nine 4×3 matrix results after the calculations. We note that in all nine matrix solutions there is only one non-zero value 1.0 that is actually corresponding to the only one resistor in the center of the circle.

$$\begin{bmatrix} 0\,0\,0\,1 \\ 0\,0\,0\,0 \\ 0\,0\,0\,0 \end{bmatrix} \begin{bmatrix} 0\,0\,0\,0 \\ 1\,0\,0\,0 \\ 0\,0\,0\,0 \end{bmatrix} \begin{bmatrix} 0\,0\,0\,0 \\ 0\,1\,0\,0 \\ 0\,0\,0\,0 \end{bmatrix} \begin{bmatrix} 0\,0\,0\,0 \\ 0\,0\,1\,0 \\ 0\,0\,0\,0 \end{bmatrix} \begin{bmatrix} 0\,0\,0\,0 \\ 0\,0\,0\,1 \\ 0\,0\,0\,0 \end{bmatrix}$$
$$\begin{bmatrix} 0\,0\,0\,0 \\ 0\,0\,0\,0 \\ 1\,0\,0\,0 \end{bmatrix} \begin{bmatrix} 1\,0\,0\,0 \\ 0\,0\,0\,0 \\ 0\,0\,0\,0 \end{bmatrix} \begin{bmatrix} 0\,1\,0\,0 \\ 0\,0\,0\,0 \\ 0\,0\,0\,0 \end{bmatrix} \begin{bmatrix} 0\,0\,1\,0 \\ 0\,0\,0\,0 \\ 0\,0\,0\,0 \end{bmatrix}$$

6 Further Discussion on the Implications of Electrical Properties on Cells and Tissues of Animals and Human Bodies

The results in Sect. 5 are based on one current injection which is not a unique result as one observes in the experiment. We now focus on the discussion regarding Eq. (6) that is derived from Finite Element (FE) method. Following many references in biology and biomedical experiments, e.g., in [6], some cells and tissues are made of water and special materials which up to present is still unknown to human beings but demonstrates many electrical properties resembling that of capacitors as well as resistors at least. In particular, the cells and tissues in different parts of the animal and human bodies vary with different circumstances such as temperatures, activity, and injecting direct or alternative current frequencies. Such electrical properties provide one with an opportunity to estimate the impedance through different aspects as long as the responses of cells and tissues to different circumstances are not simply a "linear" response.

For Eq. (6) denoted by the above formulation for the ith Alternative Current (AC) injection, suppose we inject the AC currents with different frequencies, the value $\Phi_i^{\prime} s$ should vary at least not strictly linearly by the cells and tissues' electrical properties. For each AC current injection with different frequencies,

$S_1\Phi_1 = F_1$, $S_2\Phi_2 = F_2$, ... we assume matrix S_1, S_2 ... etc remain the same that is formulated from the impedance distributions (and this assumption is rational because the impedance itself does not change much during the measurement time). Now the equations become $S\Phi_1 = F_1$, $S\Phi_2 = F_2$, ... etc. With this assumption and observation, we can have a new formulation with a much higher dimension as follows:

$$S\,[\Phi_1, \Phi_2, \ldots, \Phi_N] = [F_1, F_2, \ldots, F_N] \tag{17}$$

where N is the injection current iteration number with different frequencies and temperatures. $[\Phi_1, \Phi_2, \ldots, \Phi_N]$ is a non-singular matrix formulated by the measured voltages with different AC frequencies and temperatures.

With the proper selection of frequencies and temperatures, it is possible to obtain a more stable and unique solution to the EIT inverse problem. Further clinic empirical data is to be collected to prove the assumption and approaches however.

7 Conclusions

Although this chapter presents a discussion on the EIT algorithm development based on Finite Element (FE) method and linear algebra methods, and no general conclusions can be made, this chapter explores methods based on linear approximations for EIT inverse problem for biomedical engineering applications. At least three conclusions can be made: (1), linear approximations are possible based on Finite Element (FE) method and SVD approach, (2), it is not possible for one to uniquely determine the impedance values under only one single current injection measurement based on this SVD approach, and (3), by using electrical properties of cells and tissues of animals and human bodies, it is possible to obtain a unique and stable solution. But this depends upon the clinic data to support the electrical properties of cells and tissues. We present a discussion and a linear method in the last section on how to make use of the electrical properties in animal and human bodies by carrying out multiple measurements with different AC frequencies and temperatures.

8 Acknowledgement

The first author could appreciate the contribution of Mr. Xu wenqiang for his experiment simulations carried out while he was at Beihai College of Beihang University.

References

1. Yan, D., Jikuang, C., & Zhongkan, L. (1997). Combined variable metric method for solving electrical impedance tomography problem. *Chinese Journal of Biomedical Engineering, 16*(2), 167–173.
2. Kirsch, A. (2011). *An introduction to the mathematical theory of inverse problems* (2nd ed., Vol. 120: Applied Mathematics Sciences). New York: Springer.
3. Kirsch, A., & Grinberg, N. (2008). *The factorization method for inverse problems* (Oxford lecture series in mathematics and its applications). Oxford: Oxford University Press.

4. Rao, L., He, R., & Wang, Y. (1999). An efficient improvement of modified Newton-Raphson algorithm for electrical impedance tomography. *IEEE Transactions on Magnetics, 35*, 1562–1565.
5. Stoica, P., & Sderstrm, T. (1994). On Di's and subspace fitting approaches to direction estimation. *IEEE Transactions on Signal Processing, 42*, 691–693.
6. Riu, P. J., Rosell, J., & Pallas-Areny, R. (1992). Measurement of the frequency behaviour of tissue impedance using multifrequency tomography. In: *International Conference on Electrical Bio-impedance*.

Signal Processing for Diffuse Correlation Spectroscopy with Support Vector Regression

Peng Zhang, Zhiguo Gui, Hao Ling, Jiaxin Liu, Xiaojuan Zhang, Yiming Liu, Andi Li, and Yu Shang

Abstract Many diseases would lead to abnormal blood flow in biological tissues. In order to extract the blood flow index (BFI) in tissues with heterogeneity and irregular geometry, we previously proposed an innovative *N*th linear algorithm (i.e., NL algorithm) for technology of diffuse correlation spectroscopy (DCS), in which the DCS signals are fully utilized through iterative linear regression. With this approach, the BFI calculation is remarkably affected by the performance of linear regression. In this study, we proposed to use the support vector regression (SVR) method to denoise the DCS data by implementing the iterative linear regression. In addition, two other approaches (least-squared regression, least-absolute regression) were compared for quantitative evaluation. The DCS data generated from computer simulations with varied tissue models (i.e., flat tissue, human head, human limb, and mouse head) and those collected from the phantom experiments were utilized to evaluate the three approaches. Both simulation and phantom experiment results show that the SVR method has the best performance among three methods in extracting the BFI values, regardless of the tissue geometry and size.

Keywords Diffuse correlation spectroscopy · Blood flow index · NL algorithm · Linear regression · Denoising · Support vector regression

1 Introduction

Blood flow in biological tissues is a critical parameter for many physiological or clinical studies, as various diseases would lead to abnormal blood flow. Diffuse correlation spectroscopy (DCS) is an innovative optical technology capable of

P. Zhang · Z. Gui · H. Ling · J. Liu · X. Zhang · Y. Liu · A. Li · Y. Shang (✉)
Shanxi Provincial Key Laboratory for Biomedical Imaging and Big Data, North University of China, Taiyuan, Shanxi, China
e-mail: guizhiguo@nuc.edu.cn; yushang@nuc.edu.cn

E. T. Quinto et al. (eds.), *The Proceedings of the International Conference on Sensing and Imaging, 2018*, Lecture Notes in Electrical Engineering 606,
https://doi.org/10.1007/978-3-030-30825-4_15

noninvasive and longitudinal measurement of tissue blood flow [1–4]. Additionally, the extension of DCS technology to the diffuse correlation tomography (DCT), through which the spatial contrast of blood flow is reconstructed, has been developed over years [5–7]. DCS/DCT technologies have been extensively explored to be used in animal models and human subjects for the diagnostic and therapeutic monitoring of diseases in various tissues and organs, including brain and skeletal muscles, and tumors [3, 8, 9].

Conventionally, the blood flow index (BFI) is extracted through analytical solution or infinite element method (FEM) for DCS/DCT [4, 10]. Analytical solution generally requires a semi-infinite or layered slab geometry. Most of the biological tissues, however, are heterogeneous and with irregular geometry. FEM is unable to fully utilize the optical signals and susceptible to the data noise. In order to overcome the limitations, we proposed an *N*th-order linear (NL) algorithm [11, 12], which combines the integral form of temporal autocorrelation function ($g_1(\tau)$) with a *N*th-order Taylor polynomial. By using the NL algorithm, the temporal autocorrelation data of DCS/DCT at multiple delay times are fully utilized, and the tissue heterogeneity and irregular geometry are taken into consideration via the information of photons path lengths. The accuracy of NL algorithm had been validated through computer simulations and experimental animals [11, 12].

In practice, all the optical signals collected by the DCS instrument are mixed with noise, and signal denoising is thus an essential process in order to extract the accurate blood flow index (BFI). During the process of denoising via NL algorithm, a key step is the linear regression of autocorrelation data, from which the autocorrelation data at multiple times are fully utilized. In this chapter, we explored to use the support vector regression (SVR) method for DCS signal denoising through implementing the iterative linear regressions. Besides, the denoising outcomes from least-squared minimization (L_2 norm) and least-absolute (L_1 norm) regression approaches were compared. These approaches are evaluated and validated via computer simulations on varied tissue models (i.e., flat tissue, human head, human limb, and mouse head) as well as liquid phantom experiments.

2 Methods

In this section, we firstly describe the DCS principle and NL algorithm for BFI extraction. Then, we illustrate the linear regression approaches for DCS denoising (i.e., L_2 norm, L_1 norm, and SVR approaches). Thereafter, we elucidate the scheme of computer simulations and phantom experiments. Finally, we depict the primary methods for data analysis.

DCS Principle and NL Algorithm for BFI Extraction

The detailed description of DCS principle can be found elsewhere [4, 13, 14]. DCS flowmetry are majorly composed of a long coherent laser source at 785 nm (120 mw, Crystalaser Inc., USA), single-photon-counting detectors (Excelitas Inc., Canada), as well as digital correlator (Correlator.com, USA). The DCS source emits near-infrared light into the tissue via a multimode optical fiber with a core diameter of 100 μm (SPLF 100-3-FA, PuLei, China) located on the tissue surface. Four single-mode detector fibers with a core diameter of 4 μm (780HP, PuLei, China) were confined by a foam pad at four different separations (i.e., 1.5, 2.0, 2.5, and 3.0 cm) (Fig. 1). These detector fibers were connected, respectively, to four single-photon-counting avalanche photodiodes (APDs) inside the DCS device. The outputs of APDs were sent to an eight-channel autocorrelator board, generating normalized light intensity temporal autocorrelation functions (i.e., $g_2(\tau)$). The normalized $G_2(\tau)$ function (i.e., $g_2(\tau)$) is related to the light electric field temporal autocorrelation ($g_1(\tau)$) via the Siegert relation [4]. The value of blood flow (BFI) can be quantified because the unnormalized autocorrelation function (i.e., $G_1(\tau)$) satisfies the correlation diffusion equation. In this study, the NL algorithm proposed by us, which reflects tissue arbitrary geometry and heterogeneity, was used to extract BFI values.

NL algorithm is derived from the integral of temporal autocorrelation function $g_1(\tau)$ over individual photons, in the following expression:

$$g_1(\tau) = \frac{\langle E(0)E^*(\tau)\rangle}{\langle |E(0)|^2\rangle} = \int_0^{\infty} P(s) \exp\left(-2k_0^2 \alpha D_B s \mu_s' \tau\right) ds \quad (1)$$

Here αD_B is the blood flow index (BFI) to be extracted. τ is the delay time. $P(s)$ is the normalized distribution of detected photon path length s. k_0 is the wave

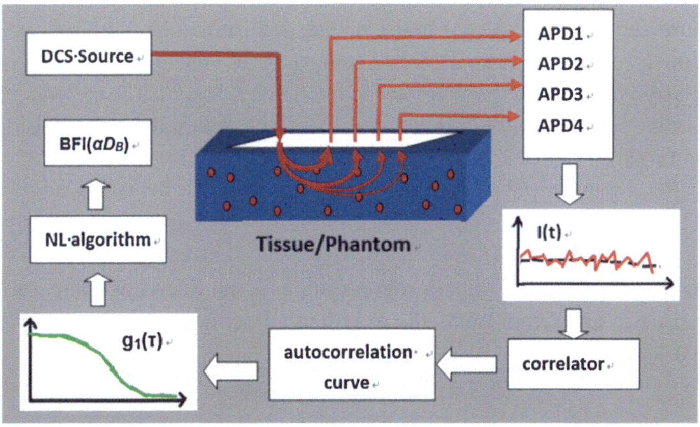

Fig. 1 DCS principle and the extraction of the blood flow index (BFI) by NL algorithm

vector magnitude of the light in the medium. When combining the Eq. (1) with the Nth-order Taylor polynomial of $g_1(\tau)$ function, and performing some mathematical derivations, we could get the following expression when τ is sufficiently small:

$$g_1(\tau) - 1 = \tau\left(H \cdot \alpha D_B^{(1)}\right) \tag{2}$$

$$\begin{aligned} g_1(\tau) - 1 - \sum_{i=2}^{N} \frac{\left(-2k_0^2\mu_s'\alpha D_B{}^{(N-1)}\right)^i \overline{s^i}}{i!}\tau^i &= \left(-2k_0^2\mu_s'\alpha D_B^{(N)}\overline{s}\right)\tau \\ &= \tau\left(H \cdot \alpha D_B^{(N)}\right), \quad N \geq 2 \end{aligned} \tag{3}$$

where

$$H = -2k_0^2\mu_s' \cdot \overline{s} \tag{4}$$

Equations (2) and (3) are referred to as the first-order and Nth-order linear algorithm, respectively. All parameters are explained in detail in [11, 15]. Once the slope (Sl) is determined from τ and $(g_1(\tau) - 1)$ data via linear regression, the unknown (αD_B in Eq. (1)) can be calculated from Eq. (2) with NL algorithm. For the Nth-order approximation ($N > 1$) in Eq. (3), which contains unknown αD_B on both sides, αD_B can be iteratively derived with the following equations:

$$g_1(\tau) - 1 - \sum_{i=2}^{N} \frac{\left(-2k_0^2\mu_s'\alpha D_B{}^{(N-1)}\right)^i \overline{s^i}}{i!}\tau^i = \tau\left(H \cdot \alpha D_B^{(N)}\right)$$

$$H \cdot \alpha D_B^{(N)} = \mathrm{Sl}^{(N)}, \quad N \geq 2 \tag{5}$$

It can be seen from Eqs. (2) and (3) that determination of the slope (Sl) via iterative linear regression is an essential step to extract the BFI (i.e., αD_B) from the autocorrelation data.

To simplify Eqs. (2) and (3) in general form, we obtain the expression of linear regression model:

$$f = kx + m \tag{6}$$

where k is the slope value of linear regression. m is the intercept. Here x represents the delay time τ, and f represents the modified autocorrelation decay (the left term on Eq. (5)).

DCS Denoising Methods

The least-squared minimization (i.e., L_2 norm) [16] is commonly used to perform the linear regression. However, the method is severely influenced by the data points far away from the regression line, leading to large errors.

In recent years, the least-absolute regression (i.e., L_1 norm) [17], which enables reducing the weight of noisy data, was adopted frequently as an alternate denoising approach for data analysis. The primary challenge for L_1 norm is how to minimize the loss function, since the loss function is non-differentiable. In this chapter, a linear regression analysis under the least-absolute deviation criterion is analyzed and transformed into a linear programming problem [15].

Very recently, the support vector regression (SVR) is emerging as an advanced approach for data linear regression [18, 19]. For general regression problems with the given training samples $D = \{(x_1, y_1), (x_2, y_2), \ldots (x_n, y_n)\}$, $x_i, y_i \in R$, we aim to obtain the value of f that is closest to y. k, m are the parameters to be determined.

The SVR problem can be expressed as follows:

$$\min_{w,b} \frac{1}{2} \|w\|^2 + c \sum_{i=1}^{m} l_\varepsilon \left(f(x_i) - y_i\right) \tag{7}$$

where l_ε is the loss function [20, 21]. When the deviations between the measured value y and the predicted value f do not exceed the tolerance of ε value, the loss value is zero. By introducing relaxation factor and Lagrange multiplier, and performing some mathematical derivations, we reach the following solution of SVR model:

$$f(x) = \sum_{i=1}^{m} \left(\hat{a}_i - a_i\right) x_i^T x + b \tag{8}$$

where

$$b = y_i + \varepsilon - \sum_{i=1}^{m} \left(\hat{a}_i - a_i\right) x_i^T x \tag{9}$$

The mathematical procedures to perform the SVR algorithm are explained in detail in [15].

Eventually, we determined the slope of the linear regression by any two points (x_i, y_i), (x_j, y_j) in the SVR model, as follows:

$$k = \frac{y_j - y_i}{x_j - x_i} \tag{10}$$

Finally, regression equation is obtained:

$$y = k \cdot x + (y_i - k \cdot x_i) \tag{11}$$

Computer Simulations

In order to evaluate the proposed SVR model, we created the following four tissue models with different size and geometries, representative of tissue heterogeneity and curvature. (1) A large slab representative of a flat tissue (Fig. 2a), (2) a large sphere representative of a human head (Fig. 2b), (3) a large cylinder representative of a human limb (Fig. 2c), and (4) a small sphere representative of a mouse head (Fig. 2d). The source-detector (S-D) separations were set as 3.0 cm for the tissues with large volumes (Fig. 2a–c) and 0.8 cm for the small mouse head (Fig. 2d). The dimensions and measurement setups for these tissue models matched approximately in vivo experiments.

In order to obtain the information of photon path length, Monte Carlo simulation of ten-million photon migrations in the tissue was utilized to generate the normalized distribution of detected photon path lengths $P(s)$ in each tissue model. The calculated $P(s)$ was then combined with the assigned BFI ($\alpha D_B = 1.2 \times 10^{-8}$ cm^2/s) as well as optical properties (e.g., μ_a and μ_s') reported in literature [13] to generate an autocorrelation function $g_1(\tau)$ based on following equation:

$$g_1(\tau) = \tau \cdot \sum_{q=1}^{Q} w(q) \exp\left(-2\sum_{i=1}^{n} k_0^2 \alpha D_B \cdot s(q) \cdot \mu_s'\right) \tag{12}$$

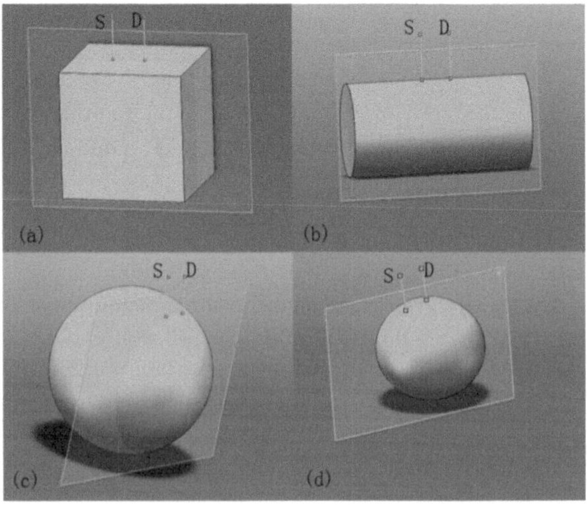

Fig. 2 Four tissue models with different sizes and geometries: (**a**) a large slab representative of flat tissues, (**b**) a large sphere representative of human head, (**c**) a large cylinder representative of human limb, and (**d**) a small sphere representative of mouse head. The separations between the source (S) and detector (D) fibers were 3.0 cm for the models of (**a**), (**b**), and (**c**), and 0.8 cm for that of (**d**). The S-D fibers were placed in the ways to mimic in vivo experimental configurations

To reflect the realistic experiments, appropriate level of noise was added to the autocorrelation function $g_1(\tau)$ at each delay time (τ). For the laser source with certain power, light intensity decreases substantially with the S-D separation, generating larger noise in DCS data. In this chapter, noise model was adopted [22, 23] to generate 1000 autocorrelation curves $g_1(\tau)$ with noise.

Phantom Experiment

In order to mimic the realistic blood flow in tissues, distilled water, India ink (Chenguang, Inc., China), and intralipid (30% solution, Huarui Inc., China) were utilized to make liquid phantoms, which provides a homogeneous tissue model. India ink was used to manipulate absorption coefficient $\mu_a(\lambda)$. The $\mu_a(\lambda)$ value was calculated based on 1% ink solution of India ink and measured with a spectrometer (QE-Pro, Ocean Optics Inc., USA). Intralipid solution was used to control the reduced scattering coefficient $\mu_s'(\lambda)$ via changing its concentration. The intralipid solution also provides the moving particles (i.e., the BFI).

An optical probe (Fig. 3b) was placed on the surface of the liquid phantom solution contained inside a rectangle glass aquarium. The custom-made holder attached to a stand holds the probe at the center of solution, in order to simulate a large slab geometry.

The optical property of liquid phantom was set initially at $\mu_a = 0.05$ cm^{-1} and $\mu_s' = 8.0$ cm^{-1}. Optical measurements were performed for 20 min and a total of

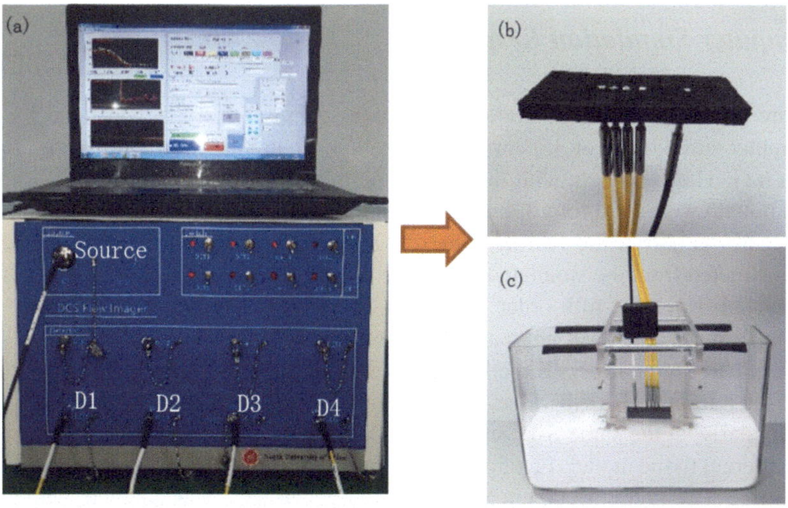

Fig. 3 (**a**) The DCS flowmetry. (**b**) A custom-made probe at the S-D separations of 1.5, 2.0, 2.5, and 3.0 cm, respectively. (**c**) Configuration for phantom experiments

1000 g_2 curves were collected by the DCS flowmetry, with sample time of 1.2 s. The averaged 50 data points at each step were used for further data analysis.

Approach Evaluation

To evaluate the three approaches (L_2 norm, L_1 norm, and SVR) in BFI extraction, we defined two variables. One variable is the error between the true and reconstructed BFIs, quantified by root mean square error (RMSE) [15]. RMSE is widely used to evaluate the error. The smaller RMSE indicates that the reconstructed BFI matches the true value. The other variable is the coefficient of variation (CV) [15], which is used to compare the robustness among the three approaches. A smaller CV value indicates a more stable status of the reconstructed blood flow.

3 Results

In this section, we present the results derived from SVR, L_1 norm, and L_2 norm approaches in computer simulations and phantom experiments. The same dataset of DCS signals were processed by each of three approaches. All the computations were carried out by a desktop PC (Lenovo ThinkCentre M8600t), with 3.4G Hz CPU and 16G memory.

Computer Simulation Result

Figure 4a shows the autocorrelation functions $g_1(\tau)$ with noise generated by the computer simulations at 3.0 cm S-D separation. Figure 4b shows the noise model for $g_1(\tau)$. The DCS data with the delay time in the range of $0.2 \leq \tau \leq 30$ μs (79 data points) were used for all linear regressions.

Figure 5a shows the RMSE values of αD_B (i.e., $\alpha D_B = 1.2 \times 10^{-8}$ cm^2/s) reconstructed from the simulated DCS data with noises. For comparisons, αD_B was calculated using the fifth-order linear model for four different tissues by the L_2 norm, L_1 norm, and SVR approaches, respectively.

As clearly seen, the SVR method performs the best in extracting the BFI values, with the minimum RMSE value.

Similarly, at the same S-D separation, SVR generated the smallest value of CV for different tissue model (Fig. 5b). Taking the human head model as an example, the CV value was 3.68% at 3.0 cm S-D separation when SVR was adopted, while this value was increased to 5.36% when the L_2 norm method was adopted, regardless of the tissue geometry (slab, sphere, cylinder) and size (human or mouse).

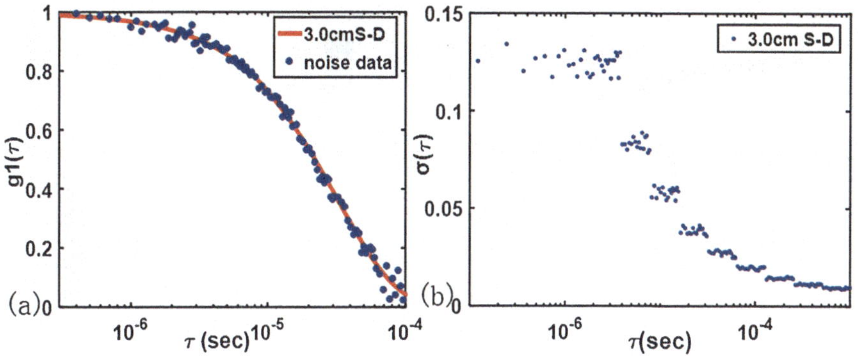

Fig. 4 (**a**) Autocorrelation functions $g_1(\tau)$ with noise generated by the computer simulations at 3.0 cm S-D separation. (**b**) The noise level of temporal autocorrelation function $g_1(\tau)$

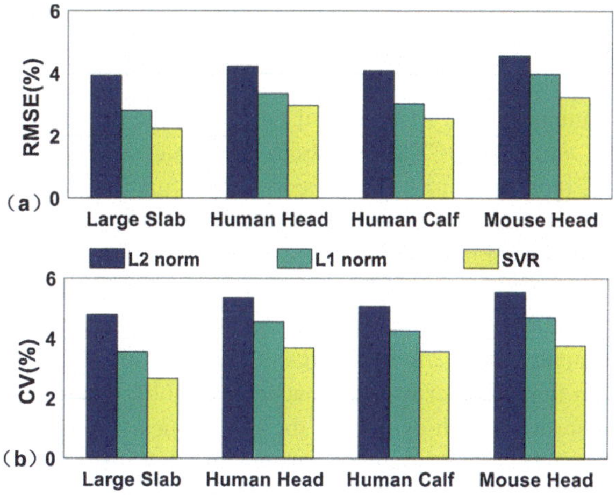

Fig. 5 (**a**) The RMSE values for four different tissue model at 3.0 cm S-D separations, when L_2 norm, L_1 norm, and SVR were used, respectively. (**b**) The CV values for four different tissue model at 3.0 cm S-D separations, when L_2 norm, L_2 norm, and SVR were used, respectively. For any of the three approaches, the NL algorithm was performed at fifth order

Additionally, we found that both error and robustness of BFI are related to the curvature of tissues. For example, the values of RMSE and CV are 2.33 and 2.66% for flat tissues (with smallest curvature) when using SVR approach. Those values are increased to 3.24% and 3.76%, respectively, for mouse head (with largest curvature). Thus, the larger curvature of tissues would lead to larger error and robustness in reconstruction of BFI value.

Phantom Experiment Result

Similar methods for data processing were applied to phantom experiments, except that the optical data were collected from liquid phantom, not from the computer

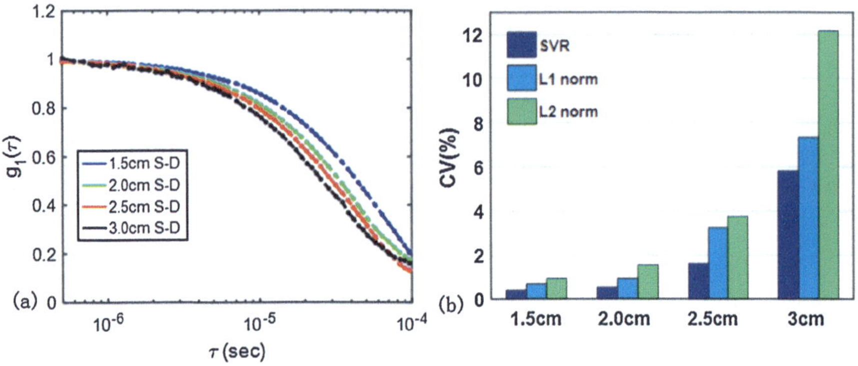

Fig. 6 (**a**) Autocorrelation functions $g_1(\tau)$ collected by the phantom experiments at four S-D separations. (**b**) The CV values for 1.5, 2.0, 2.5, and 3.0 cm S-D separations, when L_2 norm, L_1 norm, and SVR were used, respectively. For any of the three approaches, the NL algorithm was performed at fifth order

simulation. The DCS optical data from the phantom experiment match well with the simulated DCS data, in both noise intensity and distributions.

Figure 6a shows the autocorrelation function of $g_1(\tau)$ collected by the phantom experiments. As expected, the larger S-D separation leads to more noise in the correlation functions.

The CV values at 1.5, 2.0, 2.5, and 3.0 cm S-D separations derived from SVR and other approaches (i.e., L_2 norm and L_1 norm) were compared. Similar to the computer simulations, the results exhibit that the CV values by SVR are smaller than that of other two approaches (Fig. 6b), demonstrating that the SVR approach performs best in extracting the BFI values. Besides, at the largest S-D separation (i.e., 3.0 cm), average BFI values are 1.47, 1.49, and 1.51 $\times$ 10^{-8} cm^2/s by L_2 norm, L_1 norm, and SVR, respectively. These values are consistent with the reported values (0.4–2.0 $\times$ 10^{-8} cm^2/s [14]).

4 Conclusion and Discussion

In this chapter, we proposed to use SVR approach to denoise the DCS optical signals for tissues with various geometries and sizes. SVR is an advanced algorithm in category of machine learning for signal processing. For the purpose of evaluation, SVR approach was compared with the L_2 norm and L_1 norm approaches in the same DCS signals that were generated through computer simulations and phantom experiments.

The L_2 norm and L_1 norm have similar mathematical expression in the optimization problems of data fitting, but with different weight. However, both L_2 norm and L_1 norm share the limit that all data, including the noise, are involved in determining the regression line. By introducing the approach of SVR, which has been used widely for a linear or nonlinear regression, we are able to automatically remove the noisy data out of the decision plane by adjusting the penalty items. Unlike the L_2 norm, SVR does not directly exclude outliers in the data processing, hence promoting the accuracy and robustness of the regression lines. Besides, linear kernel function used in SVR would simplify the computational complexity.

In summary, the outcomes derived from both computer simulations and phantom experiments demonstrate the best performance of SVR approach in extracting the BFI values, regardless of the tissue geometry and size. Future research will focus on how to apply the SVR approach to various physiological and clinical situations.

Funding National Natural Science Foundation of China (61771433, 61671413); National Key Research and Development Program of China (2016YFC0101604); National Key Scientific Instrument and Equipment Development Project of China (2014YQ24044508); OIT Program of Shanxi Province.

References

1. Choe, R., Putt, M. E., Carlile, P. M., Durduran, T., Giammarco, J. M., Busch, D. R., Jung, K. W., et al. (2014). Optically measured microvascular blood flow contrast of malignant breast tumors. *PLoS One, 9*, e99683.
2. Chung, S. H., Feldman, M. D., Martinez, D., Kim, H., Putt, M. E., Busch, D. R., et al. (2015). Macroscopic optical physiological parameters correlate with microscopic proliferation and vessel area breast cancer signatures. *Breast Cancer Research, 17*, 72.
3. Durduran, T., Zhou, C., Edlow, B. L., Yu, G., Choe, R., Kim, M. N., et al. (2009). Transcranial optical monitoring of cerebrovascular hemodynamics in acute stroke patients. *Optics Express, 17*, 3884–3902.
4. Irwin, D., Dong, L., Shang, Y., Cheng, R., Kudrimoti, M., Stevens, S. D., et al. (2011). Influences of tissue absorption and scattering on diffuse correlation spectroscopy blood flow measurements. *Biomedical Optics Express, 2*, 1969–1985.
5. Zhou, C., Yu, G., Furuya, D., Greenberg, J. H., Yodh, A. G., & Durduran, T. (2006). Diffuse optical correlation tomography of cerebral blood flow during cortical spreading depression in rat brain. *Optics Express, 14*, 1125–1144.
6. Lin, Y., Huang, C., Irwin, D., He, L., Shang, Y., & Yu, G. (2014). Three-dimensional flow contrast imaging of deep tissue using noncontact diffuse correlation tomography. *Applied Physics Letters, 104*, 121103.
7. Zhang, X. J., Gui, Z. G., Qiao, Z. W., Liu, Y., & Shang, Y. (2018). Nth-order linear algorithm for diffuse correlation tomography. *Biomedical Optics Express, 9*, 2365–2382.
8. Shang, Y., Cheng, R., Dong, L., Ryan, S. J., Saha, S. P., & Yu, G. (2011). Cerebral monitoring during carotid endarterectomy using near-infrared diffuse optical spectroscopies and electroencephalogram. *Physics in Medicine and Biology, 56*, 3015–3032.
9. Hou, Y., Shang, Y., Cheng, R., Zhao, Y., Qin, Y., Kryscio, R., et al. (2014). Obstructive sleep apnea-hypopnea results in significant variations in cerebral hemodynamics detected by diffuse optical spectroscopies. *Physiological Measurement, 35*, 2135–2148.
10. Cheung, C., Culver, J. P., Takahashi, K., Greenberg, J. H., & Yodh, A. G. (2001). In vivo cerebrovascular measurement combining diffuse near-infrared absorption and correlation spectroscopies. *Physics in Medicine and Biology, 46*, 2053–2065.

11. Shang, Y., Li, T., Chen, L., Lin, Y., Toborek, M., & Yu, G. (2014). Extraction of diffuse correlation spectroscopy flow index by integration of Nth-order linear model with Monte Carlo simulation. *Applied Physics Letters, 104*, 193703.
12. Shang, Y., & Yu, G. (2014). A Nth-order linear algorithm for extracting diffuse correlation spectroscopy blood flow indices in heterogeneous tissues. *Applied Physics Letters, 105*, 133702.
13. Shang, Y., Chen, L., Toborek, M., & Yu, G. (2011). Diffuse optical monitoring of repeated cerebral ischemia in mice. *Optics Express, 19*, 20301–20315.
14. Huang, C., Radabaugh, J. P., Aouad, R. K., Lin, Y., Gal, T. J., Patel, A. B., et al. (2015). Noncontact diffuse optical assessment of blood flow changes in head and neck free tissue transfer flaps. *Journal of Biomedical Optics, 20*, 075008.
15. Zhang, P., Gui, Z., Guo, G. D., & Shang, Y. (2018). Approaches to denoise the diffuse optical signals for tissue blood flow measurement. *Biomedical Optics Express, 9*(12), 6170–6185.
16. Molins, A., Stufflebeam, S. M., Brown, E. N., & Hamalainen, M. S. (2008). Quantification of the benefit from integrating MEG and EEG data in minimum l(2)-norm estimation. *NeuroImage, 42*, 1069–1077.
17. Donoho, D. L. (2006). For most large underdetermined systems of linear equations the minimal l(1)-norm solution is also the sparsest solution. *Communications on Pure and Applied Mathematics, 59*, 797–829.
18. Vapnik, V. (1998). *Statistical learning theory*. New York: Wiley.
19. Vapnik, V. (1999). *The nature of statistical learning theory*. Berlin: Springer.
20. V. Cherkassky, and Y. Ma, “Practical selection of SVM parameters and noise estimation for SVM regression,” Neural Networks17, 113–126 (2004).
21. Nong, J. (2012). Parameters selection and noise estimation of SVM regression. In *Fifth International Joint Conference on Computational Sciences and Optimization* (IEEE, 2012, pp. 118–122).
22. Dong, L., He, L., Lin, Y., Shang, Y., & Yu, G. (2013). Simultaneously extracting multiple parameters via fitting one single autocorrelation function curve in diffuse correlation spectroscopy. *IEEE Transactions on Biomedical Engineering, 60*, 361–368.
23. Koppel, D. E. (1974). Statistical accuracy in fluorescence correlation spectroscopy. *Physical Review A, 10*, 1938–1945.

Intensity Inhomogeneity Image Segmentation Based on Active Contours Driven by Self-Similarity

Xu Li, Hairong Liu, and Xiaoping Yang

Abstract The segmentation of images with intensity inhomogeneity is always a challenging problem. For the segmentation of these kinds of images, traditional active contour models tend to reduce or correct the intensity inhomogeneity. In this chapter, we present a framework to make use of the intensity inhomogeneity in images to help to improve the segmentation performance. We use self-similarity measure to quantify the degree of the intensity inhomogeneity in images and incorporate it into a variational level set framework. The total energy functional of the proposed algorithm consists of three terms: a local region fitting term, an intensity inhomogeneity energy term, and a regularization term. The proposed model treats the intensity inhomogeneity in images as useful information rather than alleviates the effect of it. The proposed method is applied to segment various intensity inhomogeneous images with promising segmentation results. Comparison results also prove that the proposed method outperforms four state-of-the-art methods.

Keywords Image segmentation · Intensity inhomogeneity · Self-similarity measure · Level set · Active contour

X. Li
School of Science, Zhejiang Sci-Tech University, Hangzhou, Zhejiang, China
e-mail: lixu0103@163.com

H. Liu (✉)
School of Science, Nanjing Forestry University, Nanjing, Jiangsu, China
e-mail: hrliu@njfu.edu.cn

X. Yang
Department of Mathematics, Nanjing University, Nanjing, Jiangsu, China
e-mail: xpyang@nju.edu.cn

E. T. Quinto et al. (eds.), *The Proceedings of the International Conference on Sensing and Imaging, 2018*, Lecture Notes in Electrical Engineering 606,
https://doi.org/10.1007/978-3-030-30825-4_16

1 Introduction

Image segmentation is a fundamental task in many image processing and computer vision applications. However, due to the presence of noise, complex background, low contrast, and intensity inhomogeneity, image segmentation is still a challenging problem. In the past decades, a variety of algorithms for image segmentation have been introduced. Among them, active contour models have attracted considerable interest. The basic idea of the active contour models is to evolve the initial contour towards the object boundaries by minimizing a given energy functional. Although the energy functionals of the active contour models are diverse, they can be divided into two kinds: edge-based methods and region-based methods. Edge-based methods [1–4] guide a given contour to the object boundaries based on image gradients. The geodesic active contour model [5] is one of the most commonly used models. It utilizes the gradient information to construct an edge stopping function to stop the evolving contours on the object boundaries. Generally, edge-based approaches can provide stable segmentation results when segmenting images with strong object boundaries. However, these models suffer from the leakage problem when segmenting objects with weak boundaries. Moreover, the performance is dependent on the presence of noise as well as the position of initial contours. To overcome these problems, region-based active models have been widely studied. They use intensities or statistics like mean and standard deviation in the energy minimization frameworks. Thus, they are less sensitive to noise and initializations and perform better than edge-based active contour models for the segmentation of images with noise, intensity inhomogeneities, weak and missing boundaries. Specifically, Chan and Vese [6] simplified the Mumford–Shah [7] energy functional by using the variational level set [8] formulation and apply it to image segmentation. Suppose $I : \Omega \to R$ is the input image, C is a closed contour which can be represented by the level set function $\phi(x)$, $x \in \Omega$. The region inside/outside the contour C can be represented as $\Omega_{\text{in}} = \{x \in \Omega | \phi(x) \rangle 0\}$ and $\Omega_{\text{out}} = \{x \in \Omega | \phi(x) < 0\}$, respectively. Then the energy functional of the C-V model becomes:

$$E^{\text{cv}}(\phi, u_1, u_2) = \lambda_1 \int_\Omega (I-u_1)^2 H_\epsilon(\phi(x))\, dx + \lambda_2 \int_\Omega (I-u_2)^2 (1-H_\epsilon(\phi(x)))\, dx + \mu \int_\Omega \delta_\varepsilon(\phi(x)) |\nabla\phi(x)|\, dx,$$

where μ, λ_1, λ_2 are positive constants. u_1 and u_2 are two constants that represent the average intensities inside and outside the contour. $H_\varepsilon(\phi(x))$ is the regularized approximation of the Heaviside function defined in [8]:

$$H_\varepsilon(\phi(x)) = \begin{cases} 1 & \phi(x) > \varepsilon, \\ 0 & \phi(x) < -\varepsilon, \\ \frac{1}{2}\left\{1 + \frac{\phi}{\varepsilon} + \frac{1}{\pi}\sin\left(\frac{\pi\phi(x)}{\varepsilon}\right)\right\} & \text{otherwise.} \end{cases}$$

The derivative of $H_\varepsilon(\phi(x))$ is:

$$\delta_\varepsilon(\phi(x)) = \begin{cases} 0 & |\phi(x)| > \varepsilon, \\ \frac{1}{2\varepsilon}\left\{1 + \cos\left(\frac{\pi\phi(x)}{\varepsilon}\right)\right\} & |\phi(x)| < \varepsilon. \end{cases}$$

The C-V model assumes that the image intensity is homogeneous. However, when the image intensity is inhomogeneous, the C-V model fails to produce acceptable segmentation results. To solve the problem of segmenting intensity inhomogeneous images, a popular way is to treat the image information in local region. Li et al. proposed the active contour models: the region-scalable fitting (RSF) [9] model and the local binary fitting (LBF) [10] model which utilize the local intensity information instead of global average intensities inside and outside the contour. The energy functional of the RSF model is defined as:

$$E^{\text{rsf}}(\phi, u_1(x), u_2(x)) = \mu \int_\Omega \delta_\varepsilon(\phi(x)) |\nabla\phi(x)| \, dx + \lambda_1$$

$$\int_\Omega \int_\Omega K_\sigma(x-y)(I(y) - u_1(x))^2 H_\varepsilon(\phi(x)) \, dy dx + \lambda_2$$

$$\int_\Omega \int_\Omega K_\sigma(x-y)(I(y) - u_2(x))^2 (1 - H_\varepsilon(\phi(x))) \, dy dx$$

where μ, λ_1, λ_2 are weights of each term, $u_1(x)$, $u_2(x)$ are smooth functions to approximate the local image intensities inside and outside the contour C. $K_\sigma(y-x)$ is the Gaussian kernel function with variance σ^2 defined as:

$$K_\sigma(y-x) = \frac{1}{2\pi\sigma^2} e^{\frac{-|y-x|^2}{2\sigma^2}}.$$

K_σ has the localization property that it decreases and approaches 0 as $|y-x|$ increases. Due to the usage of local image information, these models can achieve better segmentation results than the C-V model when segmenting images with inhomogeneous intensities. Various kinds of operators that utilize local image information have been proposed to correct or reduce the effect of inhomogeneity. Zhang et al. [11] introduced a local image fitting (LIF) energy and incorporated it into the variational level set framework for the segmentation of images with intensity inhomogeneity. Wang et al. [12] proposed a local Chan–Vese model that utilizes the Gaussian convolution of the original image to describe the local image statistics. Lankton et al. [13] proposed a localized region-based active contour model, which can extract the local image information in a narrow band region. Zhang et al. [14] proposed a level set method for the segmentation of images with intensity inhomogeneity. The inhomogeneous objects are modeled as Gaussian distribution with different means and variances. Wang et al. [15] proposed an improved region-based active contour model which is based on the combination of both global and local image information (LGIF). A hybrid energy functional is defined based on a local intensity fitting term used in the RSF model and a global intensity fitting term in the C-V model. In most of the methods that can deal with intensity inhomogeneous images, the original image is modeled as a multiplicative of the bias or shading which accounts for the intensity inhomogeneity and a true image. These methods seem to produce promising segmentation results when the intensity inhomogeneity varies smoothly [16–23]. However, when the intensity inhomogeneity varies sharply (e.g., the image of a cheetah), they still cannot yield correct segmentation results. Recently, more algorithms are proposed to solve this by employing more features. Qi et al. [24] proposed an anisotropic data fitting term based on local intensity information along the evolving contours to differentiate the sub-regions. A structured gradient vector flow is incorporated into the regularization term to penalize the length of the active contour. Kim et al. [25] proposed a hybrid active contour model which incorporates the salient edge energy defined by higher order statistics on the diffusion space. Zhi et al. [26] proposed a level set based method which utilizes both saliency information and color intensity as region external energy. These models were reported effective for the segmentation of image with intensity inhomogeneity. But these kinds of features seem not powerful enough to handle images with significantly inhomogeneous intensities. More related works can be found in [27–32].

In this chapter, we propose a level set framework which can make use of the intensity inhomogeneity in images to accomplish the segmentation. Self-similarity [33] is firstly used to measure and quantify the degree of the intensity inhomogeneity in images. Then a region intensity inhomogeneity energy term is constructed based on the quantified inhomogeneity and incorporated into a variational level set framework. The total energy functional of the proposed algorithm consists of three terms: a local region fitting term, an intensity inhomogeneity energy term, and a regularization term. By integrating these three terms, the intensity inhomogeneity is converted into useful information to improve the segmentation accuracy. The proposed method has been tested on various images and the experimental results show that the proposed model effectively drives the contours to the object boundary compared to the state-of-the-art methods.

2 Model and Algorithm

Image intensity inhomogeneity exists in both medical images and natural images. Intensity inhomogeneity occurs when the intensity between adjacent pixels is different. It is observed that the pattern of intensity inhomogeneity in the same object may be similar. In other words, the intensity difference may have some continuity or consistency in the same object. That is, the quantification of intensity inhomogeneity in the same object may be homogeneous to some extent. Inspired by this, we use the self-similarity to quantify it and then incorporate it into the level set framework. In this way, the intensity inhomogeneity which is often treated as a negative effect is converted to positive effect that can help accomplish the segmentation. Figure 1 shows the flowchart of the proposed algorithm.

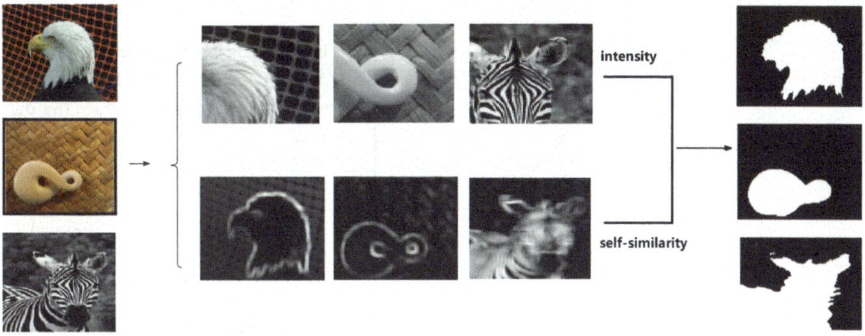

Fig. 1 The flowchart of the proposed model

Firstly, we give the definition of self-similarity. For a given $M \times M$ window W centered at x in an input image I, W can be divided into N nonoverlap small $n \times n$ patches. We denote the set of features (e.g., image intensity) in the ith small patch as F_n^i, $i = 1, \ldots, N$, then the difference between each small patch inside the window W can be defined as:

$$D_{W_x} := \{d_{i,j}\}_{N \times N} = \left\{\mathrm{Diff}\left(F_n^i, F_n^j\right)\right\}_{N \times N}.$$

Here the $\mathrm{Diff}\left(F_n^i, F_n^j\right)$ can be calculated as follows:

$$\mathrm{Diff}\left(F_n^i, F_n^j\right) = \sqrt{\sum \left(f_n^i - f_n^j\right)^2}.$$

Here f_n^i is the image intensity value in the ith patch. D_{W_x} evaluates the structure similarity inside W. It is a symmetric positive semi-definite matrix with zero-valued diagonal. If one patch is similar with the other patches inside W, then its corresponding element in D_{W_x} will be small; on the other hand, if one patch is dissimilar with the other patches, then its corresponding element in D_{W_x} will be large. Figure 2 shows the details of computing D_{W_x}. After getting the difference inside W, we can define the self-similarity measure. For a given pixel p, t_p is a template window region centered at p. The self-similarity measure (SSM) can be obtained by comparing the similarity of the D_{t_p} to D_{W_x} which is the difference of $M \times M$ window centered at x in the image. The sum of the squares differences is used to measure the similarity.

$$\mathrm{SSM}\left(t_p, W_x\right) = \sqrt{\sum_{i=1}^{N} \sum_{j=1}^{N} \left(D_{t_p}(i, j) - D_{W_x}(i, j)\right)^2}$$

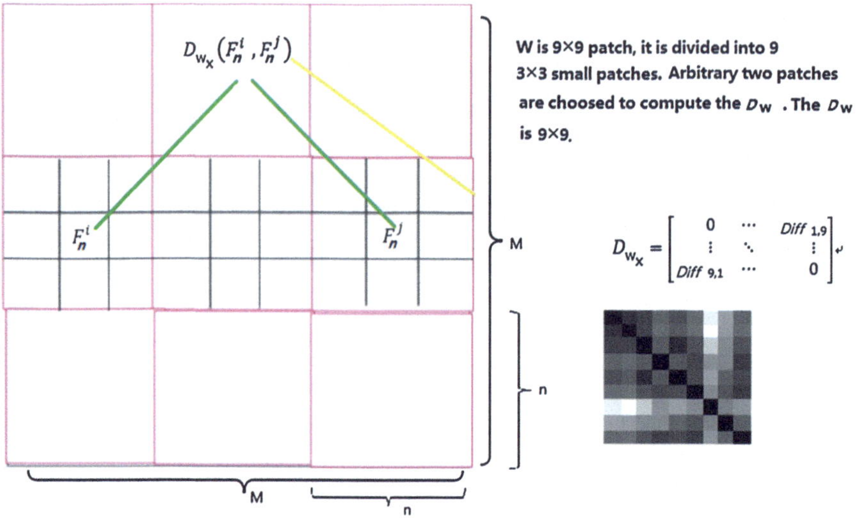

Fig. 2 The details of the computation of D_{W_x}. The D_{W_x} is a symmetric positive semi-definite matrix with zero-valued diagonal

In Fig. 3, we show an example of computing the self-similarity measure. A template region on the object is first selected. From Fig. 3, we can see that, for the region with similar intensity structure with the template region, the SSM of this region is approximately 0, and for the region with the intensity structure significantly different from the template region, the SSM of this region is bigger than 0.

In Fig. 4, more images and their corresponding SSMs are shown. We found that one can use the SSM to define a new image that describes the inhomogeneity of the original image. In this chapter, we call this image intensity inhomogeneity image. Locations with small SSM values are regarded as the same texture as the template. After we get the intensity inhomogeneity image, we can define the intensity inhomogeneity energy under the framework of C-V model:

$$E_{\mathrm{IIH}}(\phi, Ih_1, Ih_2) = \int_{\Omega_1} (\mathrm{IIH} - Ih_1)^2 dx + \int_{\Omega_2} (\mathrm{IIH} - Ih_2)^2 dx$$

$$= \int_{\Omega} (\mathrm{IIH} - Ih_1)^2 H_\varepsilon(\phi(x))\, dx + \int_{\Omega} (\mathrm{IIH} - Ih_2)^2 (1 - H_\varepsilon(\phi(x)))\, dx$$

where IIH denotes the intensity inhomogeneity image. Ω_1 and Ω_2 denote the regions inside and outside the contour, respectively. and Ih_2 are the average intensities of the IIH image inside and outside the contour.

Due to the complexity of natural images, the intensity inhomogeneity image is not powerful enough to yield an accurate result. It is necessary to also consider the intensity information of the original images. In this chapter, we use the region-scalable fitting energy in [8] to utilize the information of the original images:

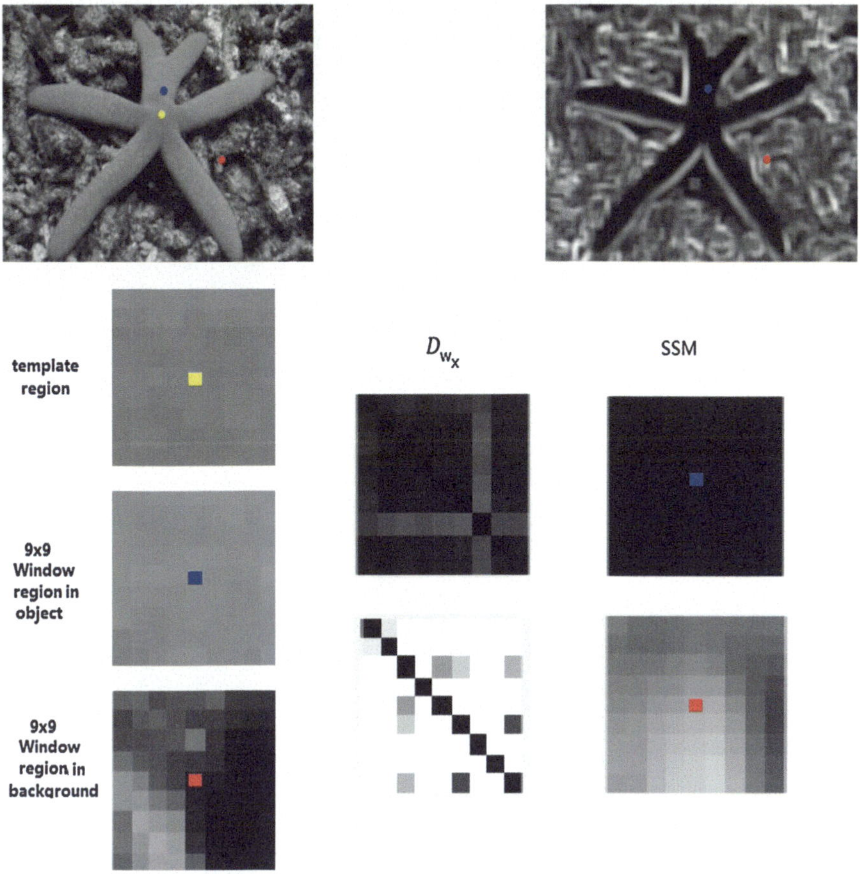

Fig. 3 One example of the SSM. Yellow point is the center of the template. Blue point and red point are two center points of the 9 × 9 region selected from object and background, respectively (The center pixels are marked as the same color in the original image). We also show the D_w inside each region and the SSM value around the center points, respectively

$$E_{\mathrm{RSF}}\left(\phi, u_1(x), u_2(x)\right) = \int_{\Omega}\int_{\Omega} K_\sigma\left(x-y\right)\left(I\left(y\right)-u_1(x)\right)^2 H_\varepsilon\left(\phi(x)\right) dydx$$
$$+\int_{\Omega}\int_{\Omega} K_\sigma\left(x-y\right)\left(I\left(y\right)-u_2(x)\right)^2\left(1-H_\varepsilon\left(\phi(x)\right)\right) dydx.$$

I is the original image. $u_1(x)$, $u_2(x)$ are smooth functions to approximate the local image intensities inside and outside the contour C. $K_\sigma(y-x)$ is the Gaussian kernel function with variation σ^2:

$$K_\sigma\left(x-y\right) = \frac{1}{2\pi\sigma^2} e^{\frac{-|x-y|^2}{2\sigma^2}}.$$

Fig. 4 The intensity inhomogeneity images (SSM) and the original images. W is a 9×9 region and it is divided into 9 nonoverlapping 3×3 small patches

By combining these energies together, we can get the following energy functional:

$$E = \lambda_1 E_{\mathrm{IIH}} + \lambda_2 E_{\mathrm{RSF}} + \lambda_3 R.$$

Here λ_1, λ_2 and λ_3 are positive constants to balance each energy term. R is the level set regularization term defined as:

$$R(\phi) = \int_{\Omega} \mid \nabla H_{\varepsilon}(\phi) \mid dx.$$

Then the total energy functional becomes:

$$E(\phi, Ih_1, Ih_2, u_1, u_2) = \lambda_1 \left(\int_\Omega (\mathrm{IIH} - Ih_1)^2 H_\varepsilon(\phi(x))\, dx \right.$$
$$\left. + \int_\Omega (\mathrm{IIH} - Ih_2)^2 (1 - H_\varepsilon(\phi(x)))\, dx \right) + \lambda_2 \left(\int_\Omega \int_\Omega K_\sigma(x-y)(I(x) - u_1(y))^2 \right.$$
$$\left. H_\varepsilon(\phi(x))\, dy dx + \int_\Omega \int_\Omega K_\sigma(x-y)(I(x) - u_2(y))^2 (1 - H_\varepsilon(\phi(x)))\, dy dx \right)$$
$$+ \lambda_3 \int_\Omega |\nabla H_\varepsilon(\phi)|\, dx.$$

Here Ih_1, Ih_2, u_1, u_2 have the following form:

$$Ih_1 = \frac{\int_{\Omega_{\text{in}}} \mathrm{IIH}\, dx}{\int_{\Omega_{\text{in}}} dx}, \quad Ih_2 = \frac{\int_{\Omega_{\text{out}}} \mathrm{IIH}\, dx}{\int_{\Omega_{\text{out}}} dx}$$

$$u_1 = \frac{\int_{\Omega_{\text{in}}} K_\sigma(x-y)\, I(y) dy}{\int_{\Omega_{\text{in}}} K_\sigma(x-y)\, dy}, \quad u_2 = \frac{\int_{\Omega_{\text{out}}} K_\sigma(x-y)\, I(y) dy}{\int_{\Omega_{\text{out}}} K_\sigma(x-y)\, dy}$$

By taking the irst variation of the energy functional with respect to ϕ, we can get the following updating equation of ϕ:

$$\frac{\partial \phi}{\partial t} = \delta_\varepsilon(\phi(x)) \left(\lambda_1 \left(-(\mathrm{IIH} - Ih_1)^2 + (\mathrm{IIH} - Ih_2)^2 \right) \right.$$
$$\left. + \lambda_2 \left(- \left(\int_\Omega K_\sigma(y-x)(I(x) - u_1(y))^2 dy + \left(\int_\Omega K_\sigma(y-x)(I(x) - u_2(y))^2 dy \right) \right) \right) \right)$$
$$+ \lambda_3 \operatorname{div} \left(\frac{\nabla \phi(x)}{|\nabla \phi(x)|} \right).$$

Different SSMs may be obtained by choosing different template regions. For example, in Fig. 5, we show different SSMs computed by using different templates. Comparing the SSMs of (b) and (c), we can see that the SSMs of (c) are more suitable to assist the segmentation. In other words, an appropriate template is very important for segmentation. In order to automatically choose the optimal position of the template, we use the following strategy: L pixels are randomly selected in the whole image and L templates can be obtained. Then we can get L SSMs by using these templates. The optimal position of the template can be selected by:

$$\min_{i \in L} \left(\frac{1}{L} \sum_{j=1}^{L} \left(\mathrm{SSM}(T_i, T_j) - \mathrm{mean}(\mathrm{SSM}(T_i, L)) \right)^2 \right)^{\frac{1}{2}},$$

Fig. 5 (**a**) The original image. (**b, c**) The SSM computed by using different template. The red dots are the center points of the template region

where $\text{mean}\,(\text{SSM}\,(T_i, L)) = \frac{1}{L}\sum_{j=1}^{L} \text{SSM}\left(T_i, T_j\right)$. This means that the template region is selected on the pixel whose standard deviation of SSM is the minimum.

3 Experimental Results

In this section, we evaluate and compare the proposed model with the C-V model [5], the RSF [9] model, the LGIF model [15], and the LSM model [14]. In these comparisons, for the C-V model, $\lambda_1 = \lambda_2 = 1$, $\Delta t = 0.1$, μ which is the weight of the regularization term is set to 6500. For the RSF model, $\lambda_1 = \lambda_2 = 1$, $\Delta t = 0.1$, $\sigma = 3$, μ is set to 5400. The LGIF model, $\lambda_1 = 0.5$, $\lambda_2 = 0.95$, they are the weights of the C-V data force and the RSF data force. $\Delta t = 0.1$, $\sigma = 3$, and the weight for the regularization term is set as $\mu = 6500$. For the LSM model, $\sigma = 3$, , the weight for the regularization term is set to 0.2. For the proposed algorithm, $N = 9$, $W = 9$, $n = 3$, $\sigma = 3$, $\lambda_1 = 4$, $\lambda_2 = 0.1$, $\lambda_3 = 5$. All the experiments are implemented with Matlab R2017b on a PC of CPU 2.8 GHz, RAM 16G.

We test the segmentation performance of the proposed model on 40 natural images with extremely inhomogeneous intensities which are collected from MSRA dataset [34] and the ECSSD dataset [35]. For quantitative analysis, we compute the F_1-measure given as:

$$F_1 = \frac{2 \times \text{Precision} \times \text{Recall}}{\text{Precision} + \text{Recall}}.$$

Here $\text{Precision} = \frac{\text{TP}}{\text{TP+FP}}$, $\text{Recall} = \frac{\text{TP}}{\text{TP+FN}}$. TP is the number of true positive pixels, FP is the number of false positive pixels, and the FN is the number of false negative pixels. In Fig. 6, we show the comparison results. Results of the quantitative

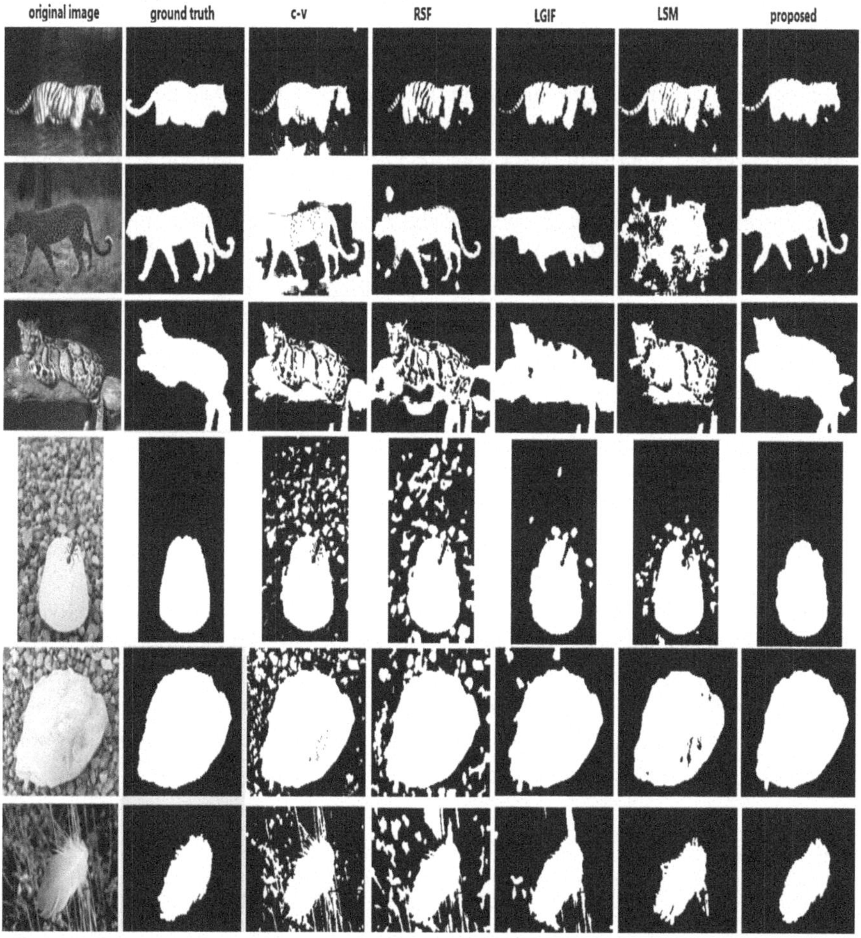

Fig. 6 Segmentation performance of each algorithm

Table 1 F_1-measure of the images in Fig. 3 and the average F_1-measure of the 40 images

	Img 1 (%)	Img 2 (%)	Img 3 (%)	Img 4 (%)	Img 5 (%)	Img 6 (%)	Average of 40 images (%)
C-V	80.2	45.7	61.9	72.7	87.8	71.1	69.2 ± 20
RSF	77	86.9	60.4	65	87.3	64.1	78.7 ± 11
LGIF	85.2	83.9	73.6	90.1	95.4	82.8	83.4 ± 9.7
LSM	87.8	74	72.8	79.2	94.7	89.1	83 ± 11
Proposed	**89.7**	**90**	**82.6**	**92.9**	**97.8**	**93.2**	**92 ± 3.9**

Black bold numbers show the efficiency of the proposed model

evaluation of these methods are shown in Table 1. In Fig. 6, the original images are shown in the first column, and the ground truths are shown in the second column. The segmentation results of the C-V, RSF, LGIF, LSM, and the proposed algorithm are shown in the other columns, respectively. In the first three images, the extremely inhomogeneous intensities are mainly in the objects and in the last three images, the backgrounds are extreme intensity inhomogeneity. It is shown that the LGIF model and the LSM model perform better than the C-V and the RSF models. In the LGIF model, the author proposed a hybrid model that combines the advantages of both global information (C-V data force) and the local intensity information (RSF data force). For the proposed algorithm, the global information is replaced by the region intensity inhomogeneity term. From the segmentation results, we can see that the proposed algorithm has the overall best performance, which further illustrates that the intensity inhomogeneity in the images can be useful information to assist the segmentation.

4 Conclusion

In this chapter, a novel active contour model is proposed for the segmentation of images with intensity inhomogeneity. We use self-similarity to quantify the intensity inhomogeneity in the images. Based on the quantified inhomogeneity, we design a region intensity inhomogeneity energy term and then incorporate it into the level set framework. The proposed model can get promising segmentation results on images with extreme intensity inhomogeneity. The experimental results show that, compared with traditional methods, the proposed segmentation model is more effective.

References

1. Kass, M., Witkin, A., & Terzopoulos, D. (1988). Snake: Active contour models. *International Journal of Computer Vision, 1*(4), 321–331.
2. Kichenassamy, S., Kumar, A., Olver, P., et al. (1995). Gradient flows and geometric active contour models., In *Proceedings of International Conference on Computer Vision* (pp. 810–815).

3. Xu, C., & Prince, J. L. (1998). Snakes, shapes, and gradient vector flow. *IEEE Transactions on Image Processing, 7*(3), 359–369.
4. Gelas, A., Bernard, O., Friboulet, D., & Prost, R. (2007). Compactly supported radial basis functions based collocation method for level-set evolution in image segmentation. *IEEE Transactions on Image Processing, 16*(7), 1873–1887.
5. Caselles, V., Kimmel, R., & Sapiro, G. (1997). Geodesic active contours. *International Journal of Computer Vision, 22*(1), 61–79.
6. Chan, T., & Vese, L. (2001). Active contours without edges. *IEEE Transactions on Image Processing, 10*(22), 266–277.
7. Mumford, D., & Shah, J. (1989). Optimal approximations of piecewise smooth functions and associated variational problems. *Communications on Pure and Applied Mathematics, 42*(5), 577–685.
8. Osher, S., & Sethian, J. A. (1998). Fronts propagating with curvature-dependent speed: algorithms based on Hamilton-Jacobi formulations. *Journal of Computational Physics, 79*(1), 12–49.
9. Li, C., Kao, C., Gore, J., & Ding, Z. (2008). Minimization of region-scalable fitting energy for image segmentation. *IEEE Transactions on Image Processing, 17*(10), 1940–1949.
10. Li, C., Kao, C., Gore, J., & Ding, Z. (2007). Implicite active contours driven by local binary fitting energy. In *Proceedings of IEEE Conference on Computer Vision and Pattern Recognition, IEEE Computer Society*, Washington, DC, USA (pp. 1–7).
11. Zhang, K., Song, H., & Zhang, L. (2010). Active contours driven by local image fitting energy. *Pattern Recognition, 43*(4), 1199–1206.
12. Wang, X., et al. (2010). An efficient local Chan-Vese model for image segmentation. *Pattern Recognition, 43*(3), 603–618.
13. Lankton, S., & Tannenbaum, A. (2008). Localizing region based active contours. *IEEE Transactions on Image Processing, 17*(11), 2029–2039.
14. Zhang, K., Zhang, L., Lam, K. M., & Zhang, D. (2016). A level set approach to image segmentation with intensity inhomogeneity. *IEEE Transactions on Cybernetics, 46*(2), 546–557.
15. Wang, L., Li, C., Sun, Q., Xia, D., & Kao, C. (2009). Active contours driven by local and global intensity fitting energy with application to brain MR image segmentation. *Computerized Medical Imaging and Graphics, 33*(7), 520–531.
16. Jiang, X., Wang, Q., He, B., et al. (2016). Robust level set image segmentation algorithm using local correntropy-based fuzzy c-means clustering with spatial constraints. *Neurocomputing*, S0925231216302351.
17. Wang, X., Min, H., Zou, L., & Zhang, Y. G. (2015). A novel level set method for image segmentation by incorporating local statistical analysis and global similarity measurement. *Pattern Recognition, 48*(1), 189–204.
18. Zhou, Y., Shi, W. R., Chen, W., Chen, Y. L., Li, Y., Tan, L. W., et al. (2015). Active contours driven by localizing region and edge-based intensity fitting energy with application to segmentation of the left ventricle in cardiac CT images. *Neurocomputing, 156*, 199–210.
19. He, C., Wang, Y., & Chen, Q. (2012). Active contours driven by weighted region-scalable fitting energy based on local entropy. *Signal Processing, 92*, 587–600.
20. Niu, S., Chen, Q., Sisternes, L., Ji, Z., Zhou, Z., & Rubin, D. (2017). Robust noise region-based active contour model via local similarity factor for image segmentation. *Pattern Recognition, 61*, 104–119.
21. Zhang, K., Zhang, L., Song, H., & Zhou, W. (2010). Active contours with selective local or global segmentation: A new formulation and level set method. *Image and Vision Computing, 28*, 668–676.
22. Li, C., Xu, C., Gui, C., & Fox, M. (2010). Distance regularized level set evolution and its application to image segmentation. *IEEE Transactions on Image Processing, 19*(12).
23. Wang, L., Chen, G., Shi, D., et al. (2018). Active contours driven by edge entropy fitting energy for image segmentation. *Signal Processing, 149*, 27–35.

24. Ge, Q., Li, C., Shao, W., et al. (2015). A hybrid active contour model with structured feature for image segmentation. *Signal Processing, 108*, 147–158.
25. Kim, W., & Kim, C. (2013). Active contours driven by the salient edge energy model. *IEEE Transactions on Image Processing, 22*, 1667–1673.
26. Zhi, X. H., & Shen, H. B. (2018). Saliency driven region-edge-based top down level set evolution reveals the asynchronous focus in image segmentation. *Pattern Recognition, 80*, 241–255.
27. Chung, G., & Vese, L. A. (2009). Image segmentation using a multilayer level-set approach. *Computing and Visualization in Science, 12*, 267–285.
28. Liu, B., Cheng, H. D., Huang, J., Tian, J., Tang, X., & Liu, J. (2010). Probability density difference-based active contour for ultrasound image segmentation. *Pattern Recognition, 43*, 2028–2042.
29. Liu, S., & Peng, Y. (2012). A local region-based Chan–Vese model for image segmentation. *Pattern Recognition, 45*(7), 2769–2779.
30. Dong, F., Chen, Z., & Wang, J. (2013). A new level set method for inhomogeneous image segmentation. *Image and Vision Computing, 31*(10), 809–822.
31. Wang, L., & Pan, C. (2014). Robust level set image segmentation via a local correntropy-based K-means clustering. *Pattern Recognition, 47*, 1917–1925.
32. Xie, X., Wang, C., Zhang, A., & Meng, X. (2014). A robust level set method based on local statistical information for noisy image segmentation. *Optik, 125*(9), 2199–2204.
33. Foote, J.T., & Cooper, M. L. (2003). Media segmentation using self-similarity decomposition. In *SPIE Proceedings, Storage and Retrieval for Media Databases* (Vol. 5021, pp. 167–175).
34. Achanta, R., Hemami, S., Estrada, F., & Susstrunk, S. (2009). Frequency-tuned saliency region detection. In *Proceedings of the IEEE Conference on Computer Vision and Pattern Recognition* (pp. 1597–1604).
35. Shi, J., Yang, Q., Li, X., et al. (2016). Hierarchical image saliency detection on extended CSSD. *IEEE Transactions on Pattern Analysis and Machine Intelligence, 38*(4), 717–729.

Application of a Temperature Measurement System Based on CCD Sensor in Blast Furnace Tuyere Raceway

Xiaoman Cheng, Shusen Cheng, Dongdong Zhou, and Ruixuan Zhang

Abstract According to the principle of spectral radiation, a color charge-coupled-device (CCD)-based pyrometer system has been established for the detection of combustion in the blast furnace tuyere raceway. The two-color method is applied to calculate the two-dimensional projection temperature distribution, as it has great advantages in the harsh measurement environment. In this chapter, the radiation images of the flame in the raceway are photographed using a color CCD camera and its two-dimensional projection temperature distribution is obtained by the two-color method and image processing technology. The results show that the average projection temperature of 320 and 5500 m^3 blast furnace are about 2260.5 K and 2266.8 K, respectively. The measured results are basically agreed with the theoretical combustion temperature. It is indicated that the temperature measurement system developed in this chapter can accurately evaluate the combustion situation in the raceway and it contributes to the stable operation of the blast furnace.

Keywords CCD sensor · Raceway · Radiation images · Two-dimensional projection temperature · Image processing

1 Introduction

The temperature field of the blast furnace raceway directly reflects the combustion behavior of the injected pulverized coal and descending coke. Observation through tuyere is the only way to monitor the inner situation of blast furnace [1]. Thermocouple is one kind of common temperature element. However, it needs to be sent into the raceway [2]. This method is difficult to implement on the blast furnace raceway. The non-contact temperature measurement method is suitable for measuring the

X. Cheng · S. Cheng (✉) · D. Zhou · R. Zhang
School of Metallurgical and Ecological Engineering, University of Science and Technology Beijing, Beijing, P. R. China
e-mail: chengsusen@metall.ustb.edu.cn

E. T. Quinto et al. (eds.), *The Proceedings of the International Conference on Sensing and Imaging, 2018*, Lecture Notes in Electrical Engineering 606,
https://doi.org/10.1007/978-3-030-30825-4_17

temperature of the raceway because of short temperature response time and non-contact [3]. Therefore, the high-temperature measurement method based on CCD image sensor has become a research hotspot in recent years.

The temperature field distribution of the flame by the area CCD is not strictly representative of a true temperature distribution, and is called the "projection temperature field" [4]. But it does not mean that measurement is not practical. F. Meriaudeau proposed a multispectral imaging system based on two CCD cameras and it should be used in the laser cladding process [5]. Huang Y et al. reported experimental investigations on temperature distributions in a 500-kW model furnace [6]. Jiang Z et al. presented a method to derive the temperature by a color CCD image for coal-fired combustion processes [7]. Temperature measurement methods based on digital imaging technology were widely used in industrial industries.

In this chapter, a pyrometer that is based on a CCD camera is used to measure the temperature distribution in blast furnace raceway. During the calibration and measurement of the blackbody furnace, it is found that the exposure time should be fixed to reduce the error, and different calibration parameters were used according to the exposure time. Finally, the experiment was carried out in two different blast furnaces. The projection temperature distribution is obtained through image processing and the two-color method.

2 Experiments

Temperature Measurement System Based on Digital Imaging

The system used in this work is shown in Fig. 1. The most important device in this system is a color CCD camera (MER-125-30UC). The sensor is Sony ICX445 CCD. The CCD sensor has Global Shutter, 1/3 in. chip, 1292 (H) $\times$ 964 (V) pixel resolution, 3.75 μm $\times$ 3.75 μm cell size. For 8 bit output, the range of gray level is from 0 to 255. The sensitivity curves of the three channels are provided by the manufacturer, as shown in Fig. 2. The focal length of lens is determined by the distance from the measured object.

Experiments are conducted on 320 and 5500 m^3 blast furnaces by this system to detect the temperature of the blast furnace raceway. The specific steps are as follows. Firstly, the lens in the peephole is checked for brightness and cleanness to avoid blocking light. Secondly, the CCD camera is installed in the front of the tuyere peephole of blast furnace to receive the light beam emitted from the raceway. Then fine-tune the camera position so that the image is as centered as possible on the screen. Finally, the camera parameters are adjusted to avoid saturation or excessively small data values for the high-temperature measurements.

The image processing procedures are shown in Fig. 3. Median filtering is a typical pre-processing step which can eliminate noise and preserve edges at the same time. The region of interest (ROI) is found by image thresholding algorithm.

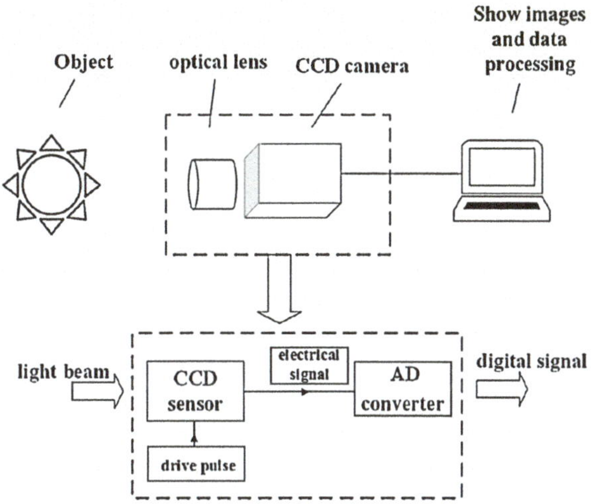

Fig. 1 Schematic diagram of the measuring system

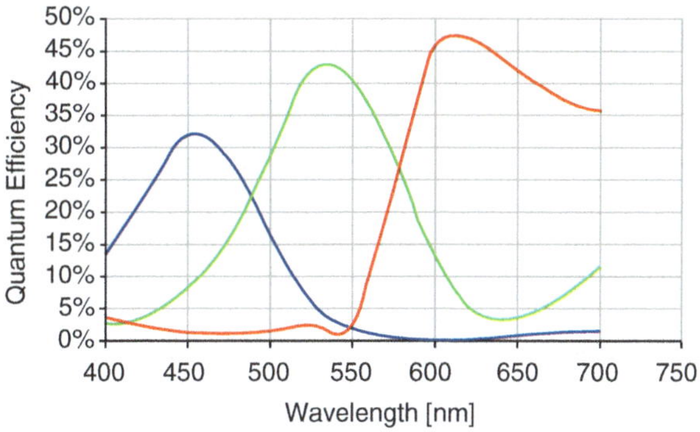

Fig. 2 Spectral response curve of CCD camera

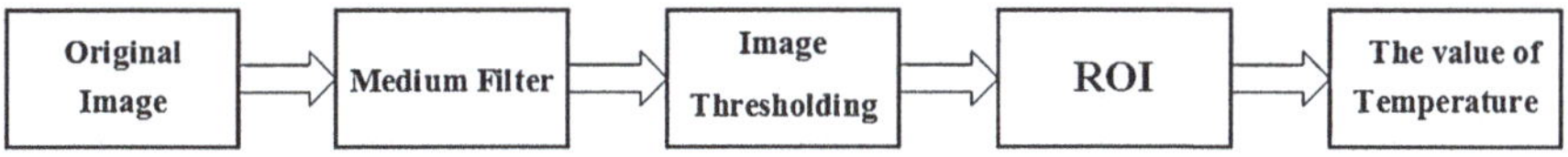

Fig. 3 Image processing routines

In this chapter, the ROI is the flame of blast furnace raceway in visible range. Then temperature can be calculated by two-color method finally.

Imaging Principle and Calculation Method

Figure 1 details the signal processing procedure. The high-temperature object emits light due to thermal radiation, and the light beam is irradiated to the CCD sensor through the optical lens. Photoelectric conversion is performed on the photosensitive pixels of the CCD sensor to form a signal charge. The signals are sequentially shifted out under the action of the drive pulse, eventually forming an electrical signal. It is converted into a digital signal via an A/D converter, and the digital signal forms a digital image stored in a memory [8]. The relationship between the gray value of the image pixel output and the spectral radiation of the measured object is given by:

$$H = \frac{\eta\mu t\omega}{F^2}\int_{\lambda_2}^{\lambda_1} \varepsilon(\lambda, T)\, E(\lambda, T)\, V(\lambda)\, d\lambda \tag{1}$$

where η is the modulus conversion coefficient, μ is the photoelectric conversion coefficient, ω is the influence coefficient of the optical system, t is the exposure time, F is the aperture value, and $\varepsilon(\lambda, T)$ is the emissivity of the object, $E(\lambda, T)$ is the monochromatic radiation power and $V(\lambda)$ is the spectral response function of the CCD camera.

Due to the spectral response function of the color CCD camera is complicated, the response curves of the red, green, and blue channels are approximated ideal, so that Eq. (1) is simplified. According to the International Commission on Illumination (CIE), the three-channel wavelengths are selected to be 700.0 nm (R), 546.1 nm (G), and 435.8 nm (B), respectively.

Defining $N_i = {}^{\eta\mu t\omega}\big/_{F^2}\, V(\lambda)$:

$$H_i = N_i \varepsilon(\lambda, T)\, E(\lambda, T)\,, i = \mathrm{R, G, B} \tag{2}$$

The two wavelengths in the two-color method can be selected, but the spectral sensitivities of the R and G channels are higher in the temperature measurement system. Moreover, it can be approximately regarded as a gray body, because when the two wavelengths are close, the flame emissivity hardly changes. The expression can be written as:

$$\frac{H_\mathrm{R}}{H_\mathrm{G}} = \frac{N_\mathrm{R}}{N_\mathrm{G}}\frac{E(\lambda_\mathrm{R}, T)}{E(\lambda_\mathrm{G}, T)} \tag{3}$$

where $N_\mathrm{R}/N_\mathrm{G}$ contains conversion coefficients and device parameters. It is necessary to be calibrated to get accurate values. Blackbody furnace as a standard reference was used to calibrate the pyrometer. The temperature range of blackbody furnace was 1773–2273 K. Images were captured at the interval of 100 K. Images with exposure times of 10, 30, 50, 70, 90, and 110 μs at the same temperature were taken.

3 Results and Discussion

Calibration in Blackbody Furnace

For images of blackbody furnace, the gray level of the two wavelengths can be obtained by using image processing technology. However, it is worth noting that when the temperature of blackbody furnace is the same, the gray value of each channel increases at different speeds with the exposure time changes, so the ratio is

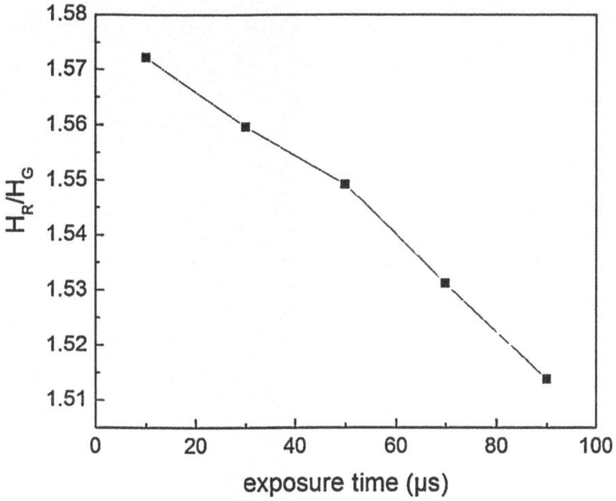

Fig. 4 The value of H_R/H_G at different exposure times

Table 1 Calculated temperature using different calibration results

The temperature of the black furnace/K	Calculated temperature/K	
	Calibration results of 10 μs	Calibration results of 70 μs
2073.0	2075.2	2034.5
2073.0	2074.1	2034.3
2073.0	2075.3	2034.8
2073.0	2074.1	2034.1
2073.0	2075.8	2035.3

different. The value of H_R/H_G at different exposure times in 2073 K are shown in Fig. 4. It can be obviously found that the ratio decreased from 1.57 to about 1.51. Theoretically, the exposure time has no effect on the value of H_R/H_G. Actually, the effect of exposure time on the pixel value is not linear, because of the loss of the signal during the imaging process [9]. So the final calibration result is not a constant.

Five images are selected randomly when the parameter is 10 μs at 2073 K. The calibration results of 10 and 70 μs were used for calculating temperature. The calculated temperature is shown in Table 1. From the comparison of the measured results, the absolute error is 2.1 and 38.4 K after averaging five images. The parameter of exposure time should be fixed, and the temperature should be calculated with the corresponding calibration results.

The Projection Temperature Distributions of Blast Furnace Raceway

The images of raceway in 320 and 5500 m^3 blast furnaces are shown in Figs. 5 and 6, respectively. There are coal lance and the pulverized coal cloud in images. These

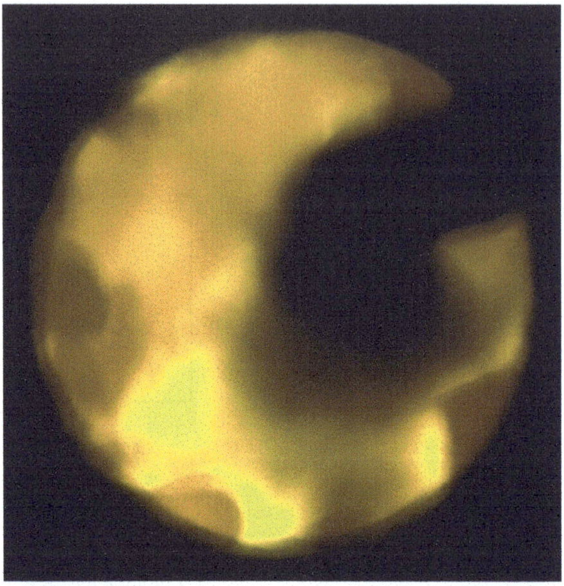

Fig. 5 The image of raceway in 320 m^3 blast furnace

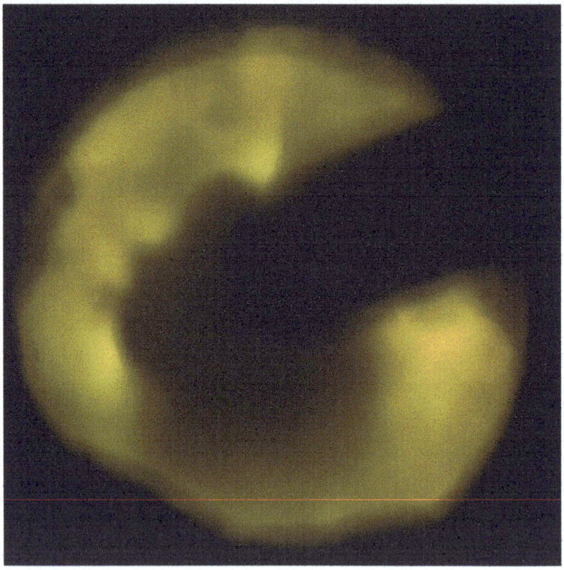

Fig. 6 The image of raceway in 5500 m^3 blast furnace

areas cannot represent the combustion, so they are removed by image processing before calculating the projected temperature field [10].

Figure 7 shows the two-dimensional projection temperature distributions of raceway of 320 m^3 blast furnace. The average temperature of the projected temperature field is about 2230.5 K. In 180 s, the temperature changed from 2205.6 to 2320.8 K. In addition, the average temperature of the raceway is 2260.5 K, and

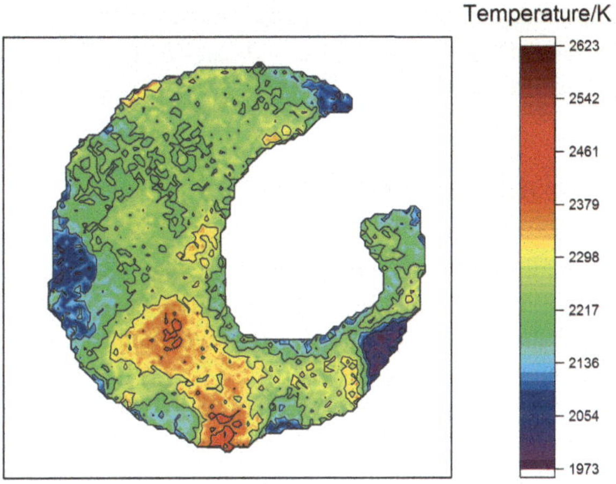

Fig. 7 The temperature distribution of raceway in 320 m^3 blast furnace

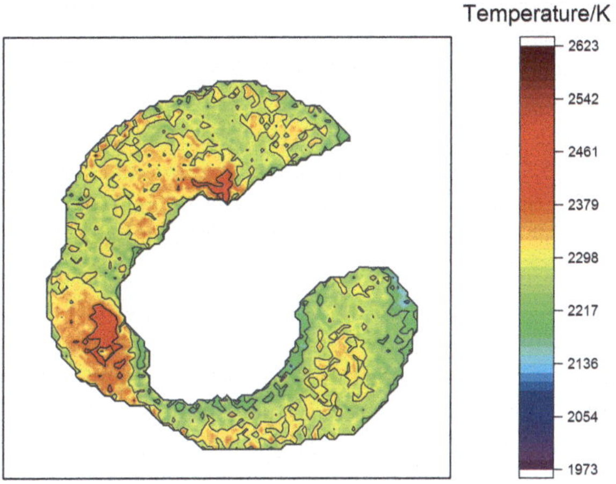

Fig. 8 The temperature distribution of raceway in 5500 m^3 blast furnace

the temperature difference is 115.2 K. The adiabatic combustion temperature of the blast furnace raceway can be a reference, and it is generally about 2273 K [11]. Due to the complex reaction before the tuyere, the theoretical combustion temperature is not the true temperature of blast furnace tuyere, but it has reference value. The measurement results are basically consistent with the adiabatic combustion temperature.

The two-dimensional projection temperature distributions of raceway of 5500 m^3 blast furnace are shown in Fig. 8. The average temperature is about 2287.2 K. The temperature changed from 2198.6 to 2384.6 K, the average temperature of raceway is 2266.8 K, and the temperature difference in 180 s is 186 K.

Comparing the images of raceway of 320 and 5500 m^3, the average temperature in 3 min of 320 m^3 is smaller than that of 5500 m^3 blast furnace. And more particles can be seen in the images of the raceway of 320 m^3 blast furnace. Two blast furnaces with different effective volumes were measured, including a small blast furnace and a large blast furnace. In previous research, 2000 and 2500 m^3 blast furnaces were successfully measured [12]. It indicates that this system can be applied to all kinds of blast furnaces.

Error Analysis and Future

To simplify the calculation of Eq. (1), the spectral response functions of the CCD camera are reduced to impulse functions approximately. It can be observed from Fig. 2 that the values of red and green channels of the CCD are obtained by the continuous spectral response.

After simplifying, the wavelength is selected according to the international regulations. Generally, the three primary colors of the CCD not necessarily use the wavelengths which they specified by CIE. However, it is necessary to consider the spectral response of the CCD. The wavelength with the largest response of the monochromatic spectrum is used as the representative wavelength, thus better reflecting the spectral intensity of the object [13].

The response values of the R and G channels are often used in the two-color method. To prevent the effects of noise and saturation, it is limited to the range of 11–245. The state of the B channel is ignored in blackbody furnaces test or industrial applications. Most of the values of B channel of images are less than 10. Ignoring the state of the B channel can interfere with the signal.

Considering the above errors, we have proposed a new optimization scheme that using a multipeak interference filter with a 3CCD camera. The filter blocks optical signals other than the designated bands [14]. There are three CCD chips in the 3CCD camera. The exposure time of the three channels can be controlled respectively [15]. The optimized system not only eliminates the above errors, but also makes the system more flexible. First, according to the temperature measurement environment and the object, the wavelength combination filter is selected. Moreover, it is possible to select the two-color method or the three-color method. In the future, we will study on the temperature measurement performance of the new system. Applications will not be limited to measuring blast furnaces, extending to other metallurgical processes such as converter steelmaking process and refining process.

4 Conclusions

The two-dimensional projection temperature measurement system based on CCD sensor was successfully established to capture the flame radiations in the tuyere raceway of 320 and 5500 m^3 blast furnaces, and the two-dimensional projection temperature distributions are gained by the two-color method.

Blackbody furnace as a standard reference was used to calibrate this system. The final calibration result is a relation between N_R/N_G and H_R/H_G. During the calibration process, parameters cannot be changed to avoid errors.

The average temperatures of images of raceway of 320 and 5500 m^3 in 3 min are 2260.5 K and 2266.8 K, respectively. The temperature of tuyere raceway changes in a short time in different blast furnaces. In the images of the raceway of smaller blast furnace, more particles can be seen.

Through error analysis, a new optimization scheme that using a multipeak interference filter with a 3CCD camera is proposed. Looking forward to more applications of CCD sensors in the metallurgical industry.

Acknowledgments This project is supported by the National Key R&D Program of China (2017YFB0304300&2017YFB0304302).

References

1. Li, Y. L., Cheng, S. S., Zhang, P., et al. (2016). Development of 3-D mathematical model of raceway size in blast furnace. *Ironmaking & Steelmaking, 43*(4), 308–315.
2. Wang, J., Wu, Z., Yan, Z. (1995). Development of temperature measuring technology for tuyere raceway of blast furnace. *Ironmaking, 14*(2), 13–16.
3. Zhang, S., Wen, L., Bai, C., et al. (2006). The temperature field digitization of radiation images in blast furnace raceway. *ISIJ International, 46*(10), 1410–1415.
4. Wei, Y. C., Yan, J. H., et al. (2002). Measurements of flame temperature distribution by the use of a colored Array CCD (charge-coupled device) (I) the measurement of a two-dimensional projection temperature field. *Journal of Engineering for Thermal Energy and Power, 17*(1), 58–61.
5. Meriaudeau, F. (2007). Real time multispectral high temperature measurement: Application to control in the industry. *Image and Vision Computing, 25*(7), 1124–1133.
6. Huang, Y., Yan, Y., & Riley, G. (2000). Vision-based measurement of temperature distribution in a 500-kW model furnace using the two-colour method. *Measurement, 28*(3), 175–183.
7. Jiang, Z. W., Luo, Z. X., & Zhou, H. C. (2009). A simple measurement method of temperature and emissivity of coal-fired flames from visible radiation image and its application in a CFB boiler furnace. *Fuel, 88*(6), 980–987.
8. Wang, Q. (2003). *Image sensor application technology*. Beijing: Publishing House of Electronics Industry.
9. Kang, L. (2009). *Establishment of the P-S furnace temperature field based on color image*. Nanchang University.
10. Zhang, R., Cheng, S., & Guo, C. (2018). Detection method for pulverized coal injection and particles in the Tuyere raceway using image processing. *ISIJ International, 58*(2), 244–252.
11. Wang, X. (2013). *Metallurgy of iron and steel (part of iron making)*. Metallurgical Industry Press.
12. Zhou, D., Cheng, S., Zhang, R., Li, Y., & Chen, T. (2018). Study of the combustion behaviour and temperature of pulverised coal in a tuyere zone of blast furnace. *Ironmaking & Steelmaking, 45*(7), 665–671.
13. Sun, Y. P., Lou, C., Jiang, Z. W., et al. (2009). Experimental research of representative wavelengths of tricolor for color CCD camera. *Journal of Huazhong University of Science and Technology, 37*(2), 108–111.
14. Fu, T., Zhao, H., Zeng, J., et al. (2010). Two-color optical charge-coupled-device–based pyrometer using a two-peak filter. *Review of Scientific Instruments, 81*(12), 124903.
15. Shao, L., Zhou, Z., Ji, W., et al. (2018). Improvement and verification of two-color pyrometry by setting exposure time respectively. *Thermal Power Generation, 47*(4), 30–36.

Weak Magnetic Stress Internal Detection Technology of Long Gas and Oil Pipelines

Bin Liu, Xiao Rui Yu, Lu Yao He, Li Jian Yang, and Yong Bin Que

Abstract The technique of weak magnetic stress internal detection is manifested in the weak magnetic field environment, including the geomagnetic field. The aim of weak magnetic stress internal detection technology is to detect the area of stress damage. Stress damage degree of ferromagnetic components is determined by detecting the weak magnetic signal on the surface of stress concentration area of ferromagnetic components. Weak magnetic stress internal detection technology of the pipeline can effectively detect the stress concentration region where the pipeline has not formed macroscopic defects, and advance the pipeline safety alarm time, which makes up for the weaknesses of the traditional magnetic flux leakage internal detection technology. In this chapter, the key technologies of the weak magnetic stress internal detection are stated. The detection mechanism, influencing factors, and engineering tests of weak magnetic stress detection technology are deeply analyzed. Furthermore, the future development direction of the weak magnetic stress internal detection technology is proposed.

Keywords Magnetic signal · Geomagnetic field · Stress concentration · Liftoff

1 Introduction

Stress concentration is an essential reason for sudden accidents in long oil and gas pipelines. In particular, a large number of stress concentration regions exist in the process of pipe manufacture and construction of new pipelines. Some stress concentration areas have reached the critical yield points, leading to the occurrence of sudden accidents after the pipeline is made available for production [1–3].

B. Liu · X. R. Yu (✉) · L. Y. He · L. J. Yang
School of Information Science and Engineering, Shenyang University of Technology, Shenyang, China

Y. B. Que
China Petroleum Pipeline Co., Ltd., West Branch, Xinjiang, China

E. T. Quinto et al. (eds.), *The Proceedings of the International Conference on Sensing and Imaging, 2018*, Lecture Notes in Electrical Engineering 606,
https://doi.org/10.1007/978-3-030-30825-4_18

In order to prevent accidents, some traditional nondestructive testing techniques are often used, including magnetic powder, magnetic flux leakage, eddy current, and infiltration and so on. These technologies all have played an important role in defect monitoring, accident prevention, and other side effects of the pipeline. However, only macroscopic defects can be found. It is impossible to carry out effective evaluation of the stress concentration region of the unformed defect due to construction, welding, foundation settlement, internal media pressure, thermal expansion, and other factors, and thus sudden accidents caused by stress damage cannot be avoided [4, 5]. From the analysis of pipeline accidents in recent years, it is obvious that new pipeline accidents happen frequently, and no macroscopic defects are formed at this time [6–8]. Weak magnetic method can effectively judge the stress concentration area of ferromagnetic metal components and evaluate the service life of ferromagnetic components. It gets the advantages of light equipment, no special magnetization, fast and convenient detection, high sensitivity, etc. Thus, this technology has been widely valued by domestic and foreign scholars [9, 10].

Based on the weak magnetic stress detection technology, this chapter first analyzes the factors that affect the weak magnetic stress detection. At the same time, the application of weak magnetic stress detection technology in engineering is summarized and the future development of this technology is prospected.

2 Weak Magnetic Stress Internal Detection Technology of Pipeline

Weak Magnetic Stress Internal Detection Mechanism of Pipeline

The weak magnetic detection method is to determine the stress damage region and the degree of stress damage of the workpiece by detecting a weak magnetic signal in the stress concentration region of the ferromagnetic metal components in the geomagnetic environment [11–14]. Under the combined action of the geomagnetic field and the applied load, the magnetic domain in stress concentration region will cause the magnetostriction. As a result, the magnetic domain lattice will be orientation and irreversible reorientation, and will form additional magnetic poles in the stress and deformation concentration region. This phenomenon macroscopic displays an abnormality of the magnetic field Hp around the component, that is the tangential magnetic signal component at the maximum stress concentration region appears an extremum $\mathrm{Hp}(x) = \mathrm{Hp}_{\max}$ and the normal magnetic signal component $\mathrm{Hp}(y)$ changes sign and has zero value in the abnormal region [15–17]. According to the ferromagnetic theory [18, 19], the stress causes change of the shape and distribution of the magnetic domain inside the ferromagnetic component and the self-magnetization increases in the form of a leakage magnetic field on the surface of the ferromagnetic component in the geomagnetic environment, especially the

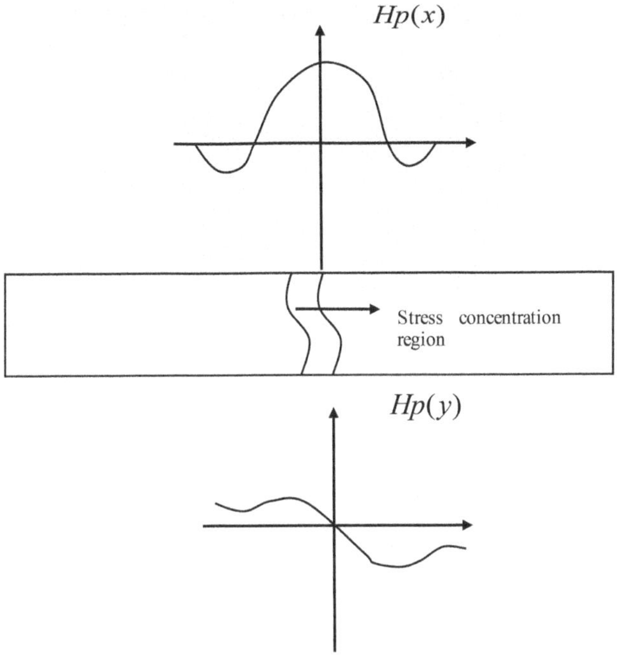

Fig. 1 Weak magnetic signal characteristics

signal normal component Hp(y) crosses zero and the tangential component reaches a maximum at the defect and stress concentration region, as shown in Fig. 1.

Weak Magnetic Stress Internal Detection Technology of the Pipeline

The pipeline weak magnetic internal detection system applies the principle of weak magnetic detection and uses the media conveyed in the pipeline as the power to complete the nondestructive testing evaluation of the pipeline. The pipeline weak magnetic internal detection system mainly includes three parts: the pipeline weak magnetic internal detection device, the mileage calibration device, and the data analysis and processing system. The structure diagram of internal weak magnetic detection device of the pipeline is shown in Fig. 2.

Pipeline weak magnetic stress internal detection device mainly realizes the detection of the stress concentration region on the pipeline and ensures the stable running of the detector. The device mainly includes four parts: a driving unit, a measuring unit, a computer unit, and a power supply unit.

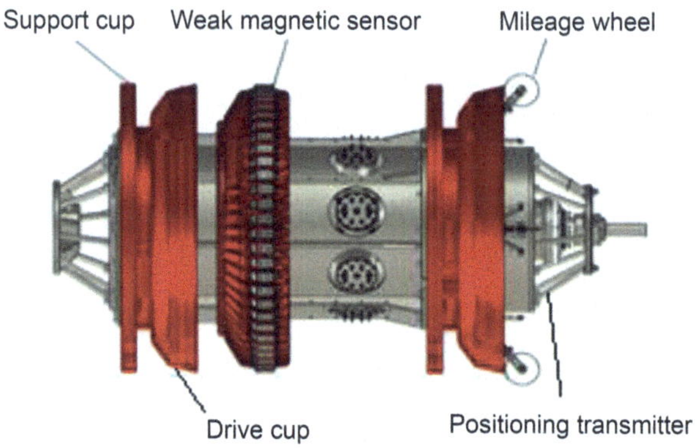

Fig. 2 Structure diagram of internal weak magnetic detection device of the pipeline

The mileage calibration device precisely locates of the stress concentration regions on the pipeline and the particular components of the pipeline. It is mainly composed of three parts: the external marking of the pipeline, the synchronous time calibration inside and outside the pipeline, and the odometer wheels record. All these three elements work together in order to record the mileage and other information.

Data analysis and processing system carry out the work of visual processing of the weak magnetic data detected by the weak magnetic sensor. At the same time, the final test results of stress concentration area of pipeline are generated. The weak magnetic sensor detects the data that are displayed on the computer after a series of processing. The data interpreter can visually observe the distribution of the stress concentration region. At the same time, the interface shows the mileage information. The location of stress concentration region is determined and marked by mileage information, providing a basis for the detection or evaluation of pipeline service life [20, 21].

3 Influencing Factors in the Internal Detection of Weak Magnetic Stress

The pipeline internal detection environment is complicated. During the detection process, there is a liftoff value between the weak magnetic stress detection probe and the pipe wall. The detection signal is influenced by external magnetic field strength, probe liftoff value, the value of the external load, and detection temperature. The external magnetic field strength and probe liftoff value has the most significant influence on the detection signal. Therefore, we focus on the analysis of the influence of the external magnetic field strength and the probe liftoff value on the weak magnetic signal in this chapter.

Influence of External Magnetic Field on Weak Magnetic Signal

In order to study the magnetic signal distribution in the stress concentration region, a magnetostatic model of the steel plate with stress concentration region is established. The size of steel plate is 470 mm × 50 mm × 14.7 mm, and the size of stress concentration region is 10 mm × 10 mm × 5 mm. According to the actual physical properties of X70 steel, the Poisson's ratio of the material properties is set to 0.28, and the elastic modulus is 2×10^6 for simulation. The magnetostatic model diagram is shown in Fig. 3.

The scanning path is set in the direction of the tensile force of 1 mm liftoff above the steel plate research area, and the variation of magnetic memory signal B above this path under different external magnetic field strength is studied. The results of the study area in the geomagnetic field environment are shown in Fig. 4. The amplitude of the magnetic signal increases and the signal characteristics become more pronounced with the increase of the stress.

Fig. 3 Steel plate magnetostatic model of the stress concentration

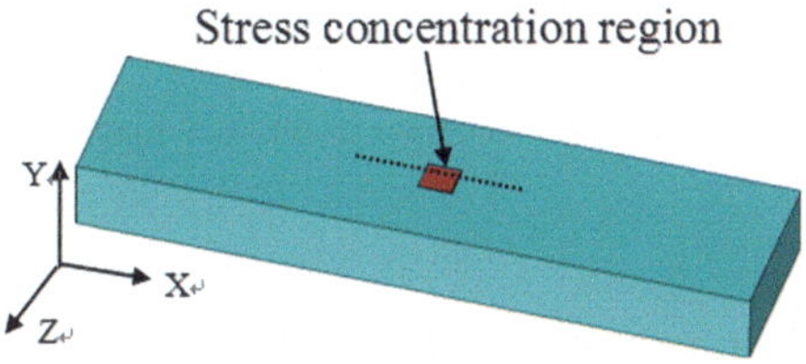

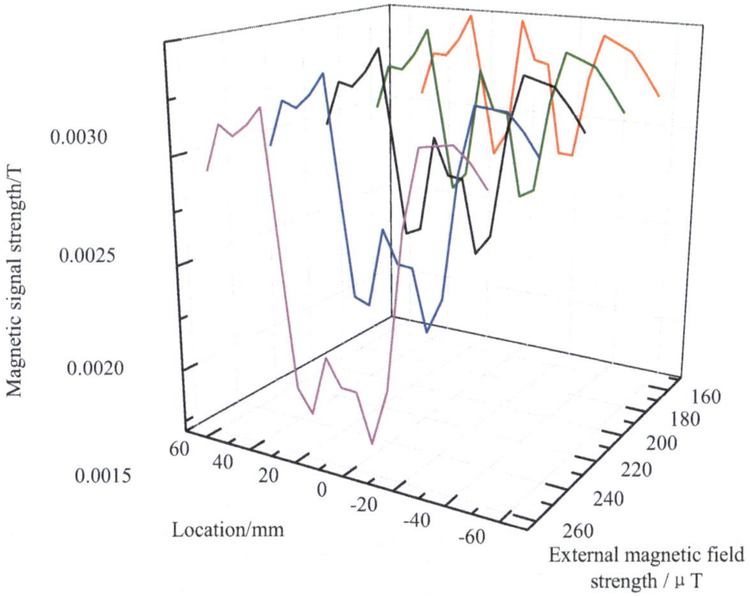

Fig. 4 Magnetic relations in the environment of the geomagnetic field

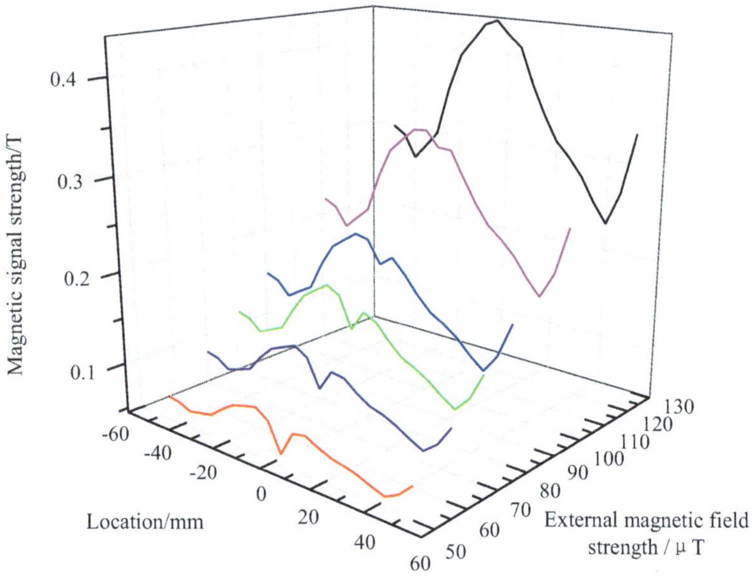

Fig. 5 Magnetics curve under different magnetic field strengths

In order to research the influence of external magnetic field on the magnetic signal in the study area, the spatial magnetic field strength should be increased in the case of the same degree of stress concentration. The magnetic curve under the action of different external magnetic fields is shown in Fig. 5. In this figure, the stress concentration region is in the range of −5 mm to 5 mm on the abscissa.

It can be seen from Fig. 5, that the value of the magnetic signal on the surface of the stress concentration area increases as the external magnetic field strength increases. However, the difference between the magnetic signal in the stress concentration region and the nearby signal reduces; when the external magnetic field reaches the maximum, the signal in the stress concentration region is unrecognizable, and the magnetic memory signal feature is covered.

Influence of Liftoff on Weak Magnetic Signal

In order to study the variation law of magnetic signal under different liftoff values, the variation law of the magnetic signal in the range of the liftoff value of 0–15 mm is shown in Fig. 6 in the case of a certain degree of stress concentration. In the figure, *X* coordinates 180–220 mm is the stress concentration region. It can be seen from Fig. 6 that there is still a magnetic signal when the measuring point has a certain liftoff value from the surface, and when the liftoff value is small, the stress concentration region can be clearly distinguished. The magnetic induction intensity

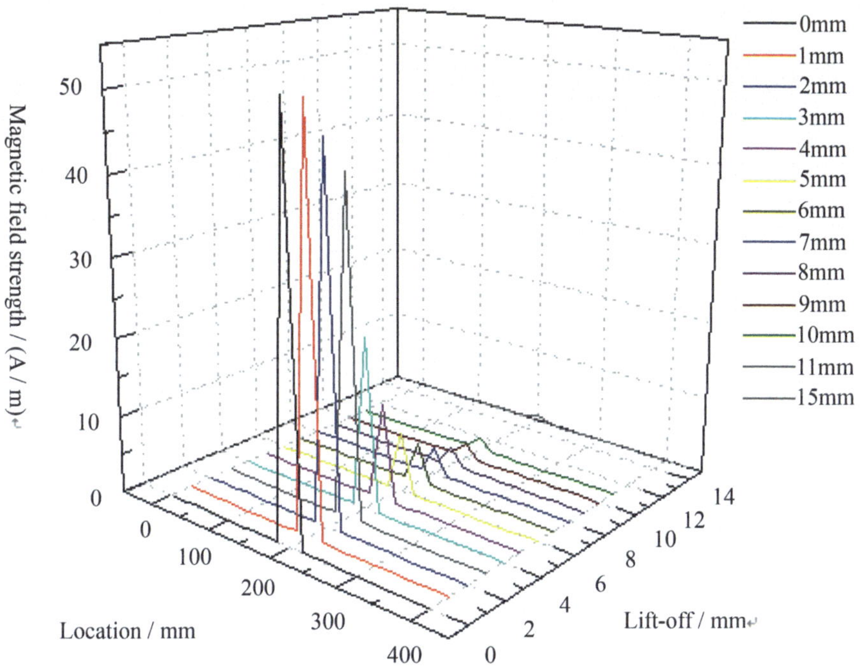

Fig. 6 The variation law of magnetic signal under different liftoff values

gradually decreases with the gradual increase of the liftoff value. When the liftoff value is less than 10 mm, it can be seen that curves calculated by the model have excellent repeatability and the stress concentration region can be identified. As the liftoff increases, the magnetic induction intensity of the steel plate surface decreases significantly, the identification level of the stress concentration zone decreases, and the magnetic signal is undetectable.

4 Engineering Application of the Weak Magnetic Stress Internal Detection Technology

In 2017, the Φ1219 weak magnetic stress internal detection equipment was developed in Shenyang University of Technology and carried out the internal testing service for the third line of the West-East Lianmuqin–Liaodun–Yandun–Hongliu–Guazhou Station in China. According to the amplitude gradient change of the weak magnetic signal, the stress concentration region level is judged. The data belongs to the normal range and detection is completed.

After the pipeline weak magnetic stress detector is operated in the pipeline, an image of the weak magnetic signal can be obtained according to the detected

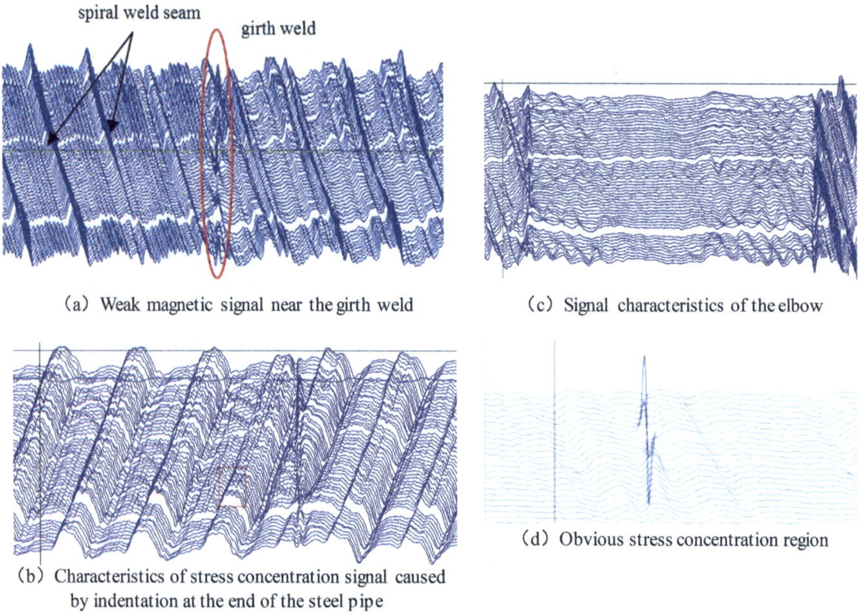

Fig. 7 Weak magnetic stress internal detection signal of the pipeline. (**a**) Weak magnetic signal near the girth weld. (**b**) Characteristics of stress concentration signal caused by indentation at the end of the steel pipe. (**c**) Signal characteristics of the elbow. (**d**) Obvious stress concentration region

amplitude of the weak magnetic signal. Magnetic memory is an internal detection signal. The signal image of the pipeline is shown in Fig. 7.

Weak magnetic stress internal detecting device is composed of a plurality of magnetic sensors as shown in Fig. 2. As shown in Fig. 7, in the detection of multichannel probes, magnetic sensor is used to detect data in each line. The signals shown in the figure have different characteristics of weak magnetic signals, which are generated by different typical characteristics of pipelines. The technique of weak magnetic inner detection is very sensitive to stress concentration region. There is a significant difference in weak magnetic signal between stress concentration region and nonstress concentration region. This makes it possible to clearly distinguish between the girth weld, the spiral weld seam, the elbow, and the like.

5 Future Prospects

At present, the weak magnetic stress internal detection technology of the pipeline has made great achievements in engineering application, however, further research is still needed in theory and engineering applications, mainly in the following three points:

1. Study on the formation mechanism of weak magnetic stress signals. It is difficult to establish the magnetomechanics coupling relationship with the traditional theory. The mechanism of weak magnetic signal generation in stress concentration region is not unified, and the mechanism research of weak magnetic stress detection technology is still a severe problem.
2. Method for judging the stress level of weak magnetic signals. The pipe forms stress concentrations under various loading conditions. When the stress concentration is at different stages, the material microstructure changes and the corresponding weak magnetic stress detection signal characteristics are different. According to this, we can effectively identify and process the weak magnetic signals at different stages and divide the stress level in different stages based on the characteristic parameters of weak magnetic signals that are the key scientific problem of component life evaluation.
3. Automatic recognition method of weak magnetic signals. At present, the identification of the internal detection data in the weak magnetic stress of long oil and gas pipelines relies on artificial recognition, which has the disadvantages of low efficiency and a large error. Realizing automatic recognition of massive data in pipeline weak magnetic stress will significantly improve data interpretation efficiency and reduce human error.

The engineering application effect of weak magnetic stress detection technology is good. At the same time, it has a broad development and application space in the field of pipeline detection. The key to the development and application of this technology is to study its basic theory.

References

1. Liu, B., He, L. Y., Huo, X. L., et al. (2017). Study on the MMM signal characteristics in magnetic field based on Kp perturbation algorithm. *Chinese Journal of Scientific Instrument, 38*(1), 151–158.
2. Liu, B., He, L. Y., Zhang, H., et al. (2017). The axial crack testing model for long distance oil-gas pipeline based on magnetic flux leakage internal inspection method. *Measurement, 103*, 275–282.
3. Gorkunov, E. S. (2015). Different remanence states and their resistance to external effects. Discussing the so-called magnetic memory method. *Insight-Non-Destructive Testing and Condition Monitoring, 57*(12), 709–717.
4. Liu, B., Cao, Y., & Wang, G. Q. (2016). Study on characteristics of magnetic memory signal of phase transition point based on the LAPW algorithm. *Chinese Journal of Scientific Instrument, 37*(8), 1825–1832.
5. Xu, K., Qiu, X., & Tian, X. (2017). Investigation of metal magnetic memory signals of welding cracks. *Journal of Nondestructive Evaluation, 36*(2), 20.
6. Jia, J. U., Feng, X. U. E., & Hong, L. I. (2015). Microstructure and magnetic property variation with addition of rare earth element Dy in Co-Ni-Al ferromagnetic shape memory alloy. *Journal of Iron and Steel Research, International, 22*(9), 858–863.
7. Liu, B., Cao, Y., Wang, D., He, L. Y., & Yang, L. J. (2017). Quantitative analysis of the magnetic memory yielding signal characteristics based on the LMTO algorithm. *Chinese Journal of Scientific Instrument, 38*(6), 1412–1420.

8. Yang, L. J., Guo, T. Z., Gao, S. X., & Liu, B. (2017). Feature analysis on magnetic flux leakage detection signals for special parts of oil and gas pipelines. *Journal of Shenyang University of Technology, 39*(01), 43–47.
9. Lo, C. C. H., Paulsen, J. A., & Jiles, D. C. (2004). Quantitative evaluation of stress distribution in magnetic materials by Barkhausen effect and magnetic hysteresis measurements. *IEEE Transactions on Magnetics, 40*(4), 2173–2175.
10. Atherton, D. L., & Szpunar, J. A. (1986). Effect of stress on magnetization and magnetostrication steel. *IEEE Transactions on Magnetics, 22*(5), 514–516.
11. Li, P., Ma, R., Chen, X., et al. (2017). First-principles prediction and Monte Carlo study of half-metallic ferromagnetism of the C-doped honeycomb CdS monolayer. *Japanese Journal of Applied Physics, 56*(4), 041201.
12. Kim, D. Y., Kim, D. H., Moon, J., et al. (2015). Determination of magnetic domain-wall types using Dzyaloshinskii–Moriya-interaction-induced domain patterns. *Applied Physics Letters, 106*(26), 190.
13. Nomura, K., & MacDonald, A. H. (2016). Quantum Hall ferromagnetism in graphene. *Physical Review Letters, 96*(25), 256602.
14. Liu, Z., Xue, F. Q., Yang, G. G., et al. (2013). Experimental measurement method of sensitivity matrix for electromagnetic tomography. *Chinese Journal of Scientific Instrument, 34*(9), 1982–1988.
15. Ren, J. L., Sun, J. L., Zhou, P., et al. (2013). Engineering application prospect and research on magnetic memory in two-dimensional testing. *China Academic Journal Electronic Publishing House, 49*(22), 8–15.
16. Sun, X., & Liu, B. (2014). Weak magnetic detection for stolen oil hole on long-distance transportation pipeline. *Journal of Shenyang University of Technology, 36*(04), 436–440.
17. Wang, X., Wang, D. S., Wu, R., et al. (2016). Validity of the force theorem for magnetocrystalline anisotropy. *Journal of Magnetism & Magnetic Materials, 159*(3), 337–341.
18. Wang, D. S., Wu, R., & Freeman, A. J. (2015). Magnetocrystalline anisotropy of interfaces: First-principles theory for Co-Cu interface and interpretation by an effective ligand interaction model. *Journal of Magnetism & Magnetic Materials, 129*(2–3), 237–258.
19. Hillebrands, B., & Ounadjela, K. (2015). Spin dynamics in confined magnetic structures. *Applied Physics Letters, 72*(18), 159–724.
20. Yang, L. J., Shen, W., Gao, S. Z., & Liu, B. (2018). Low frequency electromagnetic tracking and positioning technology for pipeline internal detector. *Journal of Shenyang University of Technology, 40*(01), 48–53.
21. Yang, L. J., Ma, F. M., & Gao, S. W. (2016). Research of on-line inspecting pipeline magnetic flux leakage system. *Chinese Journal of Scientific Instrument, 37*(8), 1763–1780.

Index

A
Active contours
  edge energy, 188
  localized region-based, 187
  models, 186
Adaptive division, 54, 57–58, 60
Algorithm performance analysis
  clutter cancellation analysis, 81–82
  data and experiment parameters, 80–81
  optimal moving target detection, 82–85
  SCNR improvement, 81–82
Along-track interferometry (ATI), 78
Approximated Gaussian, 33–34
Attribute reduction
  FRCMbR, 7–8
  fuzzy rough clustering, 6–7
  of rough set theory, 3
Azimuth aperture analysis
  DPCA and STAP, 77–78
  method, 79–80
  processing chain, 78–79

B
Behavior recognition
  color feature extraction (*see* Color moment)
  feature extraction, 116
  sketch-based systems, 116
Biomedical applications, 164
Blast furnace
  injected pulverized coal, 199
  projection temperature distributions, 203–206
  ROI, 201
  temperature distribution, 200
Blood flow index (BFI)
  approach evaluation, 180
  computer simulations, 178–181
  DCS, 173
    denoising methods, 177–178
    principle, 175–176
  FEM, 174
  and NL algorithm, 175–176
  optical signals, 174
  phantom experiment, 179–182

C
Calibration, 137–138
CCD sensor
  calculation method, 201–202
  calibration in blackbody furnace, 202–203
  digital imaging, 200–2001
  error analysis, 206
  imaging principle, 201–202
  projection temperature distributions, 203–206
  temperature
    field distribution, 200
    measurement system, 200–2001
    thermocouple, 199
Circular-SAR (CSAR), 80, 86
Clustering
  algorithm, 2
  analysis, 1
  attribute reduction (*see* Attribute reduction)

E. T. Quinto et al. (eds.), *The Proceedings of the International Conference on Sensing and Imaging, 2018*, Lecture Notes in Electrical Engineering 606,
https://doi.org/10.1007/978-3-030-30825-4

Clustering (*cont.*)
FRCMbR, 8–12
fuzzy rough algorithm, 2, 6–8
RCM, 5–6
Color moment
classification, 117–118
contour feature extractions, 117–120
robustness, 118
segmentation, 117–118
Compact polarimetric SAR (CP SAR), 88
Constant false alarm rate (CFAR) detector
detector, 100–103
post-target discrimination processing, 83
radiometric adjustment, 78
ship detection experiments, 98
Convolution network
anisotropic Gaussian kernel, 125
Chan–Vese model, 187
fast RCNN, 54
Gaussian functions, 32
non-maximum suppression, 46
and pooling, 55, 59
Coupled stripline resonator, 152

D
Deep learning, 46, 54
Denoising, 46, 117, 174, 177–178
Diffuse correlation spectroscopy
and DCT, 174
denoising methods, 177–178
innovative optical technology, 173
principle, 175–176
signal denoising, 174
signals, 182
Displaced phase center array (DPCA), 77

E
Edge detection, 131–133
Electrical impedance tomography (EIT)
in applied mathematics, 164
FE, 165–166
Maxwell differential equations, 164–165
method, 163–164
SVD (*see* Singular value decomposition (SVD))
Electrical resistance tomography (ERT)
conductivity distribution, 16
dual-plane, 74
horizontal slurry flow (*see* Horizontal settling slurry flow)
image reconstruction analysis, 18–20
online measurement, 66
principle of, 16–17
reconstruction image, 16
sensitivity coefficient amplitude distribution, 17–18
system and principle, 67–70

F
Feature extraction
of athletes, 116
and classification, 53
color (*see* Color moment)
motion trajectory, 122–125
physical difference, 88
of players, 118–120
stadium line, 120–121
UAV flight image, 55
Finite element (FE), 17, 18, 20, 69, 164–166, 169, 170, 174
Flow visualization
high-performance ERT system, 74
horizontal settling slurry (*see* Horizontal settling slurry flow)
Football game
color component, 119
trajectory generation diagram, 124
video segment, 123
Frequency mixing
cavity resonators, 145
digital means, 133
left resonator, 135
magnetic field, 131, 149
microprocessor, 139
operation, 137
oscillator, 133, 134
probe, 148
Q-factor, 152
sampling, 35
spatial, 146
temperature and moisture, 153
UWB radars, 156
Fuzzy c-mean clustering (FCM), 2, 4–5, 9
Fuzzy rough clustering based on attribute reduction (FRCMbR) algorithm
experimental setting, 8–10
parameter experiments, 10–12
Fuzzy theory
clustering analysis, 4
FCM, 4–5
nonlinear problems, 2
RCM, 5–6
and rough set theory, 2, 6, 7, 12

G
Gaussian smoothing approximation, 33–34
Geomagnetic field, 213
Ground range detected high resolution (GRDH), 98

H
Horizontal settling slurry flow
  characteristics, 67
  ERT system and principle, 67–70
  experiment setup, 70–71
  gravity effect, 66
  phase
    velocity, 66
    volume fraction, 66
  regimes, 66–67
  results and discussions, 71–73
Hough transform, 120, 121, 128
Hyperspectral image (HSI)
  data, 110
  grid-structured 2D array, 105
  IID methods, 106
  physical to imaging model, 107–108
  reflectance recovery, 108–110
  results and analysis, 111–112
  shading component, 106
  surface topographical structures, 107

I
Image processing, 42, 47, 50, 117, 200, 204
  computer vision applications, 186
  intelligence mining system, 42
  photogrammetry, 47
  routines, 201
  UAV reconnaissance information (*see* Unmanned aerial vehicle (UAV))
Image segmentation, 186
Induction sensors, 140
Intelligence support, 42
Intensity inhomogeneity
  and algorithm, 188–194
  C-V model, 187–188
  edge-based active contour models, 186
  experimental results, 194–196
  Heaviside function, 186–187
  image segmentation, 186
  model, 188–194
Internal electrode excitation
  ERT system, 20, 21
  image reconstruction, 20–23
  mixed mode, 23–25
Intrinsic image decomposition (IID), 105–106

K
Key-point detection, 36–37
  DOG detector, 36
  repeatability, 38
  SIFT scale space construction, 32–33

L
Level set, 186–188, 196
LC oscillators, 133, 135–137
Liftoff
  tensile force, 213
  weak magnets
    signal, 214–215
    stress detection probe, 212
Linear regression, 174, 176, 177, 180
Local solid velocity, 66

M
Magnetic induction, 214, 215
Magnetic signal
  engineering application, 215–216
  external magnetic field, 213–214
  internal detection technology, 210–212
  liftoff, 214–215
  macroscopic defects, 210
  stress concentration, 209
Maxwell differential equations, 164–165
Microwave microscopy, 145–149
Microwave sensors
  active/passive radiometry, 157–158
  dimensional measurements, 145–146
  high-frequency electromagnetic waves, 145
  microscopy, 146–149
  NDT (*see* Nondestructive testing (NDT))
  resonant frequency, 145
  testing, 144
Mixed excitation, 23–25
mm-wave sensors, 213–215
Moisture content, 152–153

N
NL algorithm, 174–176, 181, 182
Nondestructive testing (NDT)
  magnetic internal detection system, 211
  techniques, 210
  testing materials, 143

O
Object detection algorithm
  building training set, 59
  faster RCNN network architecture, 55, 56

Object detection algorithm (*cont.*)
labeling, 59
loss function
classification, 57
RPN, 56–57
network
training, 59, 60
validation, 59–61
reconnaissance images, 60, 62–63
target, 53
Operating frequency, 137

P
Performance, 140
Polarization features, 88, 91
Projection temperature field, 200, 203–206

R
Raceway
blast furnace, 199
projection temperature distributions, 203–206
Radar
antenna, 89
methods, 155–157
NDT (*see* Nondestructive testing (NDT))
pulsed, 154
ship detection algorithms, 88
Radiation images, 144, 157, 201, 202, 206
Reconnaissance information mining system
equipment, 43
faster RCNN network, 54
images, 60–62
methods, 41
mission system, 43
photoelectric payload, 44
UAV (*see* Unmanned aerial vehicle (UAV))
Reconstruction algorithm, 24, 67, 68
Reflectance
data, 110
IID methods, 106
low-frequency geometry cues, 106
physical to imaging model, 107–108
quantities, physical, 107
recovery, 108–109
results and analysis, 111–112
shading component, 106
surface normals, 109–110
Reflection, 90, 145, 148, 154–155
Repeated averaging filters
approximate Gaussian filtering, 32
Gaussian filtering, 39
image, 34
scale space construction, 34–36
simulation and experiment, 148
Resonance
configuration of coils, 134
ferromagnetic, 148
induction methods, 132
metal strip edge detection, 132–133
mixer method, 133–135
proposed method, 135–136
sensor, 133
Resonant testing methods, 150–152
Rough c-mean clustering (RCM), 5–6
Rough set theory, 2–4, 6, 12

S
Scale space construction, 34–36
Scattering testing methods, 88, 90, 91, 154–155
Sea clutter distribution
CFAR detector, 100
detection performance, 102–103
distribution model, 99–100
evaluation methodology, 100–101
goodness-of-fit analysis, 101–102
histogram, 100
and ships, 88
Self-similarity measure, 188–190, 196
Sensitivity matrix, 17, 18
Sentinel-1 IW level-1, 98–99
Ship detection
algorithms, 88
backscattering signal, 97–98
CFAR detector, 98, 102
circular polarization ratio, 91–92
comparative analyses, 98
decompositions, 88
relative phase, 93–94
roundness, 92, 93
SIFT key-point detection
applications, 31
DOG blob detector, 32
experimental results, 37–38
extraction, 44
space construction, 32–33
Signal-to-clutter noise ratio (SCNR), 78–81, 85

Simulation
computer, 178–181
conductivity distribution models, 25
and experiment, 148
experimental conditions, 25–26
GPR scanning, 156
Monte Carlo, 178
with noise, 27–28
noise free, 26, 27
Singular value decomposition (SVD)
approximations, 164
EIT inverse problem, 166–167
electrical properties, 169–170
experiments, 167–169
Space-time adaptive processing (STAP), 77
Statistical distribution, 102, 116
Stress concentration
ferromagnetic metal components, 210
magnetic curve, 214
magnetostatic model, 213
mileage calibration device, 212
pipe manufacture and construction, 209
region level, 215
tangential component, 211
Support vector regression (SVR), 174, 177, 180–183
Surface normals, 109–110
shading component, 106
single real-world HSI, 106
Switches, 138
Synthetic aperture radar (SAR), 43, 88, 97
application, 77
CP, 88
detection results, 48
dual polarimetric, 88
imaging, 80, 98
marine management, 88
ocean security, 88
radar system, 88
single and dual polarimetric, 88
single-channel, 78

T

Temporal and spatial interest points, 125–127
Trajectory growth method
feature extractions, 122–125
high-level recognition algorithm, 128
spatial relationship, 116
Transmission, 144–146, 151, 152, 154–155
Two-dimensional projection temperature, 121, 149, 204–206

U

Unmanned aerial vehicle (UAV)
artillery correction, 49–50
damage assessment, 49–50
electronic map fusion, 43–44
image mosaic, 43–44
infrared image fusion, 44–46
object detection (*see* Object detection algorithm)
participation in the war, 42
reconnaissance methods, 41
requirements analysis, 42–43
target
detection and recognition, 46, 47
location, 47–49
visual, 44–46

X

X-Bragg scattering model, 90, 91

The manufacturer's authorised representative in the EU is Springer Nature Customer Service Centre GmbH, Europaplatz 3, 69115 Heidelberg, Germany. If you have any concerns regarding our products, please contact ProductSafety@springernature.com

Printed and bound by CPI Group (UK) Ltd, Croydon, CR0 4YY
30/09/2026
02225651-0004